79th Annual Meeting of the Northeastern Weed Science Society 2025

Proceedings – Volume 79

Annapolis, Maryland, USA
6-9 January 2025

Editor:

Vijay Singh

ISBN: 979-8-3313-2755-2

Printed from e-media with permission by:

Curran Associates, Inc.
57 Morehouse Lane
Red Hook, NY 12571

Some format issues inherent in the e-media version may also appear in this print version.

Northeastern Weed Science Society (NEWSS)
P.O. Box 25
Woodstown, NJ 08098
USA

Phone: 315-209-7580

northeasternweedscience@hotmail.com

Additional copies of this publication are available from:

Curran Associates, Inc.
57 Morehouse Lane
Red Hook, NY 12571 USA
Phone: 845-758-0400
Fax: 845-758-2633
Email: curran@proceedings.com
Web: www.proceedings.com

PROCEEDINGS - VOLUME 79

79th Annual Meeting of the Northeastern Weed Science Society (NEWSS)

The winning image in the 2025 NEWSS Photo Contest, submitted by Claudio Rubione, Virginia Tech

Site of 2025 annual meeting:

Annapolis, MD

January 6-9, 2025

The Graduate Hotel

Proceedings
of the
Seventy Ninth Annual Meeting
of the
Northeastern Weed Science Society

Vijay Singh, Editor

NORTHEASTERN
WEED SCIENCE SOCIETY

2025 SUSTAINING MEMBERS

PLATINUM

BASF
We create chemistry

Bayer CropScience

CORTEVA agriscience

syngenta

SILVER

AMVAC
An American Vanguard Company

FMC

pbi/gordon companies
Employee-Owned

BRONZE

GDM

LABServices
The life-science technology support team

Nufarm

Sustaining members: https://www.newss.org/about/sustaining-members/

Table of Content

† Represents student contest, and * represents presenter

NEWSS EXECUTIVE COMMITTEE – OFFICERS

President **Erin Hitchner**
Syngenta
erin.hitchner@syngenta.com

President-elect **Thierry Besancon**
Rutgers University
thierry.besancon@rutgers.edu

Vice President **Katelyn Venner**
BASF
Katelyn.venner@basf.com

Treasurer **Larissa Smith**
Syngenta
larissa.smith@syngenta.com

Past President **Wesley Everman**
North Carolina State University
wes_everman@ncsu.edu

NEWSS COMMITTEES – CHAIRS

Editor	Vijay Singh	Virginia Tech
Membership Chair	Matt Elmore	Rutgers University
Public Relations	John Kaminski	Penn State University
Research and Education	Carolyn Lowry	Penn State University
Sustaining Membership Weed Contest	Caio Brunharo	Penn State University
CAST Representative	Anthony Witcher	Tennessee State University
Graduate Student Rep	Eli Russell	Virginia Tech
WSSA Representative	Steve Pyle	Syngenta
Media Communication	Lynn Sosnoskie	Cornell University
Science Policy Director	Lee vanWychen	WSSA

NEWSS COMMITTEES – CHAIRS

Student Paper Contest	Jacob Barney	Virginia Tech
Student Poster Contest	Anthony Witcher	Tennessee State University
Photo Judging Chair	Hilary Sandler	University of Massachusetts
Site Selection Committee	Thierry Besancon	Rutgers University
Awards Committee	Wesley Everman	North Carolina State University
Nomination Committee	Dwight Lingenfelter	Penn State University
Resolutions Committee	Hilary Sandler	University of Massachusetts
Collegiate Weed Contest Committee	Caio Brunharo	Penn State University
Audit Committee	Larissa Smith	Syngenta
Archives Committee	Art Gover	Penn State University
Past Presidents Committee	Wesley Everman	North Carolina State University
Endowment Committee	Wesley Everman	North Carolina State University

2025 NEWSS PROGRAM

Program Committee	Thierry Besancon (Chair)
	Katelyn Venner
	Carolyn Lowry
	Vijay Singh
	Eli Russell
Section Chairs	
Agronomy	Ramon Leon
Weed Ecology and Invasive Species	Sarah Kezar
Turf, Ornamental, and Non-Crop	David Hillger
Research Posters	Kate Venner
Fruits, Vegetables, and Tree Crops	Roger Batts

2025 NEWSS AWARDS

NEWSS Fellow Award

Carroll Moseley
Head of State Regulatory Affairs & Stewardship
Syngenta (retired)

Outstanding Researcher Award

John Wallace
Associate Professor of Weed Science
Pennsylvania State University

Garry Schnappinger Service Recognition Award

Steve Pyle
Syngenta

SOAR Award from IR4

Mark VanGessel
Professor and Extension Specialist
University of Delaware

Robert D. Sweet Outstanding Graduate Student – Ph.D.

Eli Russel
Virginia Tech University

Robert D. Sweet Outstanding Graduate Student – M.S.

Jackson Alsdorf
North Carolina State University

2025 NEWSS CONTEST WINNERS

Poster Winners

First Place	Juan Romero, Virginia Tech
Second Place	Navdeep Godara, Virginia Tech
Third Place	Jenna Beville, Virginia Tech

Oral Competition Winners

First Place	Navdeep Godara, Virginia Tech
Second Place	Laurel Wellman, Penn State University
Third Place (Tie)	Hunter Lee, North Carolina State University
	Juan Romero, Virginia Tech
Honorable mention	Kylie Bill, Cornell University

Graduate Student Writing Contest

First Place (PhD)	Aleah Butler-Jones, Cornell University
Second Place (PhD)	Preetman Bajwa, Cornell University
First Place (MS)	Henrique Scatena, Cornell University

Photo Contest

First Place	Claudio Rubione, USDA-GROW
Second Place	Laurel Wellman, Penn State University
Third Place (Tie)	Juan Romero Cubas & Eli Russell, Virginia Tech

1

Impact of Commercial Surfactants on Soil Mobility and Environmental Leaching of Residual Herbicides

W. Bouchelle[1], L. Foster Seybert[2], L. Sosnoskie[2], C. Duplais[2], T. Besançon

[1]Rutgers University, New Brunswick, United States, [2]Cornell University, Geneva, United States, [2]Cornell University, Geneva, United States

Herbicides play a critical role in various crop production systems. The availability of spray surfactant products has increased significantly, with chemical supply companies promoting them for purposes such as drift control, spreading, and binding enhancement. This study evaluated the efficacy of six spray surfactants in combination with two commonly used herbicides. With the increasing frequency of intense rainfall events, such as heavy thunderstorms, maintaining herbicide efficacy has become a significant challenge for farmers. This trial simulated a heavy rainfall event to assess whether the inclusion of surfactants in the spray mixture improves herbicide binding to the soil surface. Results indicate that without activation water, pendimethalin concentrations in the soil were lower compared to treatments with activation water, suggesting rapid photodegradation under sunlight exposure. Pendimethalin showed minimal response to surfactant addition and remained resistant to water treatment effects. A significant difference was observed between the herbicide-treated control without irrigation and all other treatments. In the case of atrazine, a significant decrease in concentration was noted due to rapid soil degradation and the combined effects of water treatment and leaching. In atrazine-treated soils, the non-irrigated control exhibited higher pre-treatment concentrations compared to most other groups, except those treated with the surfactants Grounded and NU Film P. Five weeks after treatment, the addition of Grounded, NU Film, and Induce significantly reduced weed counts compared to atrazine alone without irrigation. All herbicide treatments resulted in significantly fewer weeds than the untreated irrigated and non-irrigated controls. In conclusion, pendimethalin demonstrated resistance to water treatment, whereas atrazine concentrations were significantly impacted by leaching and degradation. Surfactants did not substantially influence pendimethalin dynamics but may affect the movement and efficacy of atrazine, particularly in mitigating leaching. These findings highlight the importance of herbicide solubility and activation water in influencing environmental mobility. Moreover, surfactants like Grounded and NU Film P show potential to improve atrazine performance. Future research should further explore the interactions between surfactants, herbicides, and soil properties, including variations in soil texture, such as differing proportions of sand, silt, and clay, to better understand their effects on herbicide efficacy and environmental behavior.

2

Confirmation of Glyphosate- and Atrazine-Resistant Palmer amaranth (Amaranthus palmeri) Populations in New York and New Jersey and Responses to Alternative Chemistries.

A. Butler-Jones[1], E. Maloney[1], W. Kramer[2], S. Morran[3], T. Gaines[2], T. Besançon[4], L. Sosnoskie[1]

[1]Cornell University, Geneva, United States, [2]Colorado State University, Fort Collins, United States, [3]Grains Research and Development Corporation, Barton, Australia, [4]Rutgers University, Chatsworth, United States

Palmer amaranth (*Amaranthus palmeri* S. Watson) is one of the most troublesome weeds in North America due to its rapid growth rate, substantial seed production, competitiveness, and the evolution of herbicide-resistant populations. Though native to the southwestern United States and northern Mexico, *A. palmeri* is found throughout the US including New York and New Jersey. Between 2019 and 2022, three NY populations (NY-GEN, NY-ORA, and NY-STE) were discovered in soybean [*Glycine max* (L.) Merr.] rotations following ineffective applications of glyphosate. In 2019, a NJ population was collected in a soybean field at the Rutgers Agricultural Research & Extension Center (RAREC) in 2019. This research, conducted from 2023 to 2024, aimed to (1) describe the response of *A. palmeri* populations from NY and NJ to glyphosate and atrazine through dose–response studies and (2) assess the sensitivity of the NY populations to alternative herbicide SOAs used in northeastern agronomic and specialty crops. Based on the effective dose necessary to reduce aboveground biomass by 50% (ED_{50}), the NY populations were 42 to 67 times more resistant to glyphosate compared with a glyphosate-susceptible population. Additionally, we confirmed the presence of glyphosate-susceptible individuals in NJ; however, the range in glyphosate sensitivity within this population warrants further investigation to determine the degree of susceptibility. The presence of atrazine-resistant *A. palmeri* in NY (NY-GEN, NY-STE) was also confirmed. The NY *A. palmeri* populations were sensitive to label rate applications of alternative chemistries except for the ALS inhibiting herbicides; however, additional research is needed to confirm resistance to chlorimuron, chloransulam, and other Weed Science Society of America (WSSA) group 2 chemistries. To date, the NY populations were effectively controlled by label rate applications of herbicides belonging to WSSA groups 4, 10, 14, and 22. Additional research is warranted to characterize the susceptibility of these populations to herbicides within other WSSA groups and to develop effective *A. palmeri* management strategies suitable for northeastern crop production.

3

Industrial Hemp (Cannabis sativa L.) and Weed Responses to Biostimulants

M. Bohrer[1], A. Thomas[1], J. Petzoldt[1], L. Sosnoskie[1]

[1]*Cornell AgriTech, Geneva, United States*

Hemp (*Cannabis sativa* L.) is a versatile crop grown for fiber, grain, and cannabinoids. The 2018 U.S. Farm Bill spurred a significant expansion in hemp production. Early establishment and weed control are critical, as competitive weeds can cause significant yield losses. Currently, only Sonalan HFP is labeled for use in hemp; as such, growers should ensure that crop growth and vigor is maximized. To improve crop competitiveness, growers may consider using biostimulants, although information is lacking. In 2024, greenhouse trials examined how hemp ('Bialobrzeskie' a monoecious, dual purpose hemp variety known for high quality fibers and seed production) and weeds [Palmer amaranth (*Amaranthus palmeri* S.Wats) and velvetleaf (*Abutilon theophrasti* Medik.) respond to two biostimulants: gibberellic acid (GA) and a seaweed extract.

The treatments included a plant hormone extract, Gibberellic Acid 90% from Power Grown, with a recommended 1x dose of 100 ppm, and a seaweed extract, Vitazyme from Vital Earth with a recommended 1x dose of 912 l ha^{-1}. Foliar applications for both products were administered at rates of 0x (nontreated check), 0.25x, 0.5x, 1x, 2x, and 4x. In run 1 (October 2024), 14 replicates of each product-dose combination were applied to hemp plants at the 6–8 leaf stage. In run 2 (November 2024), 4 replicates of each product-dose combination were applied to hemp and each weed species at the 2–4 leaf stage. Applications were made using a cabinet sprayer set to deliver 187 l ha^{-1}. Plant height and leaf count were measured at 5, 11, and 15 days after treatment (DAT). Aboveground fresh biomass was recorded at 15 DAT ; dry biomass estimates were collected 4-7 days later. The greenhouse trials were conducted using a randomized complete block design.

Hemp treated with GA responded positively to increasing doses. Concentrations of 1-4x GA doubled hemp heights (40 cm in run 1, 20-30 cm in run 2) relative to the nontreated check (20 cm in run 1, 10 cm in run 2). Vitazyme did not have a similar effect on hemp heights (approximately 20 cm across all doses in round 1 and 10 cm in round 2). Neither product increased hemp leaf number nor dry biomass at harvest under the conditions of this study. Averaged over doses, Palmer amaranth treated with GA averaged 0.05 g more in dry biomass compared to plants treated with Vitazyme. No significant differences were observed between products with respect to Palmer amaranth leaf number and height. Averaged over doses, velvetleaf treated with GA produced significantly more leaves compared to velvetleaf treated with Vitazyme. There were no significant differences between products with respect to velvetleaf height and dry biomass. Future iterations of this preliminary study will be conducted in the spring and summer under more optimal growing conditions. Additional biostimulants will be included for comparison. We will also investigate co-planting weeds and crop plants to more closely observe weed/crop competitive interactions as mediated by biostimulants. Future studies will also scale up promising treatments to field trials.

4

Herbicide-Resistance In Italian Ryegrass Populations From Delaware and Maryland

A. Veron[1], B. Scott[2], M. Vangessel[2], K. Vollmer[3], C. Brunharo[1]

[1]Penn State, University Park, United States, [2]University of Delaware, Georgetown, United States, [3]University of Maryland, Queenstown, United States, [1]Penn State, University Park, United States

Italian ryegrass (*Lolium multiflorum*) presence in Delaware and Maryland has become more problematic in recent years. An increase in small grain infestations has resulted in yield loss and grain quality issues with ryegrass seed as a contaminant. There has also been an increase in questions concerning lack of control in no-till corn. Herbicide-resistant Italian ryegrass was first documented in 1998 for Group 1 herbicides and in 2012 for Group 2 herbicides. Multiple-resistance had not been confirmed in either state or there has been no data demonstrating the extent of Italian ryegrass resistance in the region. During the spring of 2023, fields with mature Italian ryegrass were identified and seedheads collected. Forty-one sites had enough viable seed to include in this trial. Resistance was tested for glyphosate, pyroxsulam, nicosulfuron, pinoxaden and clethodim.

None of the accessions were resistant to glyphosate or clethodim; but all were resistant to pyroxsulam. Twenty-five accessions were resistant to both pyroxsulam and nicosulfuron. Five accessions were resistant to pyroxsulam and nicosulfuron and pinoxaden. In small grains, farmers cannot rely on Group 2 herbicides for Italian ryegrass control. In addition, careful consideration is needed with using Group 1 herbicides and implement resistance management strategies to reduce the risk of selecting for additional resistance.

5

Tank Mixing 2,4-D, Fomesafen, and S-metolachlor for Weed Control in Enlist Soybean.

K. Vollmer[1]

[1]*University of Maryland, Queenstown, United States*

Herbicide spray programs should aim to abate herbicide resistance development. Tank mixing different herbicide sites-of-action is one tactic to delay the onset of herbicide-resistance. 2,4-D-choline and fomesafen are two herbicides that can be used in the Enlist® soybean system. Furthermore, data on optimal application timing of this mixture is needed to both improve herbicide-resistance management and reduce herbicide application costs. Herbicide labels recommend spraying 2,4-D and fomesafen when weeds are less than 15 cm, but 2,4-D will not provide additional residual weed control, leading to the possibility of an additional postemergence application. A study was conducted in 2024 at the University of Maryland Wye Research and Education Center to 1) evaluate the efficacy of 2,4-D + fomesafen tank mixtures for weed control, and 2) evaluate the efficacy of early and late POST applications for weed control in Enlist® soybean. The study was a two-factor factorial arranged in a randomized complete block design with 4 replications. Factors consisted of herbicide application timing and herbicide treatment. The entire study area was conventionally tilled prior to planting, and a preemergence application of *S*-metolachlor (1,068 g ha^{-1}) was applied at planting. Postemergence herbicide applications were made either 2 (EPOST) or 4 (LPOST) weeks after planting and consisted of fomesafen (421 g ha^{-1}), fomesafen (421 g ha^{-1}) + *S*-metolachlor (1,602 g ha^{-1}), fomesafen (421 g ha^{-1}) + 2,4-D (1,065 g ha^{-1}), or fomesafen (421 g ha^{-1}) + 2,4-D (1,065 g ha^{-1}) + *S*-metolachlor (1,602 g ha^{-1}). A nonionic surfactant (0.25% v/v) was included with these treatments. A glyphosate (1,104 g ae ha^{-1}) + ammonium sulfate (1.9 kg ha^{-1}) check was included for comparison and applied with the EPOST treatment. Application timing did not affect common lambsquarters or morningglory control, but it did affect giant foxtail control 7 weeks after soybean planting. 2,4-D + fomesafen or 2,4-D + fomesafen + *S*-metolachlor controlled common lambsquarters (*Chenopodium album* L.) and morningglory (*Ipomoea* spp.) at least 98% compared to fomesafen alone and fomesafen + *S*-metolachlor (67%). Applications made EPOST controlled giant foxtail (*Setaria faberi* Herrm.) 70% compared to applications made LPOST (39%) regardless of herbicide treatment. Average soybean yield was 3,026 kg ha^{-1}, but there were no differences in application timing or herbicide treatment. These results show 2,4-D + fomesafen to be a more effective combination for controlling broadleaf weed species compared to fomesafen alone. Both fomesafen and *S*-metolachlor can provide some residual grass activity. Better giant foxtail control with EPOST treatments could have been the result of overlapping residual control from the initial *S*-metolachlor treatment. However, additional research is needed to corroborate these results and determine the effectiveness of these mixtures on other weed species.

Gap in pagination due to unavailable paper.

Page 6

7

Herbicide Resistant Weeds in the Mid-Atlantic Region of the United State

A. Thomas[1], J. Petzoldt[1], M. Bohrer[1], L. Sosnoskie[1]

[1]*Cornell AgriTech, Geneva, United States*

Herbicides are highly effective tools for managing weeds that are widely used, globally. However, their extensive and consistent use exerts strong selective pressure on weeds, driving the evolution of herbicide resistance. Documenting cases of herbicide resistance is critical for several reasons. Firstly, accurate documentation helps stakeholders understand which herbicides are no longer effective in controlling specific weed species. This knowledge is essential for selecting alternative weed management strategies and avoiding the use of ineffective treatments, which saves time and resources. By identifying resistant populations early, targeted actions can be taken to slow the spread of resistance. This includes rotating herbicides with different modes of action, integrating non-chemical methods, and implementing best practices to prolong the efficacy of existing herbicides. When herbicides fail, farmers may resort to overusing chemical controls or using more toxic options, potentially harming non-target organisms and ecosystems. Documentation supports sustainable practices by promoting integrated weed management (IWM) strategies that minimize environmental harm. Finally, tracking resistance patterns provides data for researchers studying the mechanisms of resistance and helps guide the development of new herbicides or alternative control methods. Understanding resistance trends can also inform future research priorities. This poster summarizes the distribution of herbicide resistance as of 2024 across the Mid-Atlantic and New England regions. The study area includes Connecticut (CT), Delaware (DE), Massachusetts (MA), Maryland (MD), Maine (ME), New Hampshire (NH), New Jersey (NJ), New York (NY), Pennsylvania (PA), Rhode Island (RI), Virginia (VA), Vermont (VT), and West Virginia (WV). Data were compiled from *The International Herbicide-Resistant Weed Database* (weedscience.org) and correspondence with regional weed scientists.

In total, >20 weed species (including Palmer amaranth (*Amaranthus palmeri* S. Wats) and waterhemp (*Amaranthus tuberculatus* (Moq.) J.D.Sauer), among others) in the Mid-Atlantic and New England regions have evolved resistance to six modes of action (MOAs). Among them, at least 10 species have developed resistance to multiple MOAs. Notably:

Common ragweed (*Ambrosia artemisiifolia* L.) exhibits resistance to four MOAs (Groups 2, 5, 9, and 14), the highest among all species. Common lambsquarters (*Chenopodium album* L.) shows resistance to Group 5 herbicides, reported in almost every state. Resistance to Group 2 herbicides is the most prevalent, with at least 14 species across the region exhibiting this trait. At least 9 instances of multiple resistance have been reported across species.

Copies of this poster will be available through Cornell University. Please contact Dr. Lynn Sosnoskie (lms438@cornell.edu) for more information or to suggest corrections and/or additions.

8
Tulipalins: A Natural Herbicide from a Tulip Bulb Waste Stream.

J. Peters[1], M. Elmore[1]

[1]*Rutgers University, New Brunswick, United States*

There is interest among the public to reduce synthetic herbicide use, but few practical alternatives are available. The only currently recognized organic pre-emergent is corn gluten which has limited efficacy. Preliminary screening indicates that tulipalin which is a lactone found in tulip bulbs has herbicidal properties both as a pre-emergent and as a post-emergent.

9

Determining the Impact of a Microbial Soil Amendment on Cover Crop Residue Degradation to Promote Anaerobic Soil Disinfestation

T. Campbell[1], M. Cutulle[1]

[1]*Clemson University, Charleston, United States*

Initially developed to supplant conventional fumigation practices, anerobic soil disinfestation (ASD) is a soil sterilization technique which utilizes the byproducts of existing anaerobic soil microbes to control various soil born pests. A carbon source is incorporated into the soil, then covered with a tarp or plastic mulch, and finally saturated with water. Consumption of carbon material by anaerobes, and the subsequent production of byproducts such as various volatile organic compounds (VOC), is temperature and time dependent. ASD typically requires four weeks during warmer periods and six weeks during cooler periods to see desired effects. The goal of this research is to determine if a soil microbial amendment helps facilitate the breakdown and consumption of carbon amendments. This study took place in a climate-controlled greenhouse (night/day temperatures ranging from 18-23° C) at Clemson University's Coastal Research and Education Center (CREC) (32.794196° N, -80.067747° W) in the fall of 2024 and was designed as a randomized complete block design with three replications. Each experimental unit consisted of a single 2.5-gallon bucket filled with a consistently heaping amount of soil gathered from an on-site certified organic field. The treatments were as follows: no cover crop + no soil microbial amendment, cover crop (crimson clover at 33.63 kg ha^{-1}) + no soil microbial amendment, and cover crop + soil microbial amendment (14.03 L ha^{-1}). To quantify the facilitation and breakdown of carbon amendments data loggers were used to measure soil temperature, pH, and ORP (oxidation redox potential) with the final aim of using these measurements to calculate cumulative anaerobicity over an approximately four-week period. Other data gathered includes number of instances where yellow nutsedge (*Cyperus esculentus* L.) punctured through the plastic mulch cover, as well as yellow nutsedge, broadleaf, and total weed germination counts post-ASD once the plastic mulch was removed. When averaged across replications, treatments containing cover crop residue tended towards higher cumulative anaerobicity values than treatments containing no cover crop, while the addition of the soil microbial amendment further increased values. Instances of yellow nutsedge puncturing the plastic mulch were relatively low across all treatments (one in treatment one and one in treatment three). Across replications, total weed counts post-ASD were numerically lower in the treatment containing the soil microbial amendment (11 vs 15 and 16, respectively).

10

Barriers and Solutions to On-Farm Wildflower Strip Establishment: A Survey of USA Growers.

L. Smith[1], A. Ditommaso[1], J. Losey[1]

[1]*Cornell University, Ithaca, United States*

On-farm wildflower strips have many benefits, both for growers and for the environment; however, low adoption of the practice indicates large barriers to wildflower strip establishment. We carried out a survey of 266 US growers to assess their attitudes towards the practice of on-farm wildflower strips and what they see as the major barriers to establishing on-farm wildflower strips. We also surveyed interest in a seed pelleting technology that would make it possible to plant wildflower strips using a conventional field crop planter. Responses indicated that many growers would like to establish wildflower strips, and are prevented from doing so primarily by labor, cost, and equipment restraints. These results indicate that a technology to reduce the labor and cost of planting wildflower strips could increase the adoption of on-farm wildflower strips in the US.

11

Control of Succulent Pea (Pisum sativum) in Snap Bean (Phaseolus vulgaris)

L. Sosnoskie[1]

[1]*Cornell AgriTech, Geneva, United States*

In 2021, volunteer garden pea (*Pisum sativum* L.) plants were frequently observed in commercial snap bean (*Phaseolus vulgaris* L.) fields, prompting growers to seek management strategies. However, controlling volunteer peas in snap beans is challenging due to limited herbicide options, as both crops are legumes and share multiple registered chemistries. Reflex® (POST) and Sandea® (PRE and POST) are not registered for peas, and were the primary herbicides considered for volunteer pea suppression by growers. A trial in 2022 at Cornell AgriTech in Geneva, NY, evaluated the efficacy of labeled herbicides for volunteer pea control and their impact on snap bean yields. Volunteer peas ('Little Marvel') were seeded at a mean density of 10 plants per 7.6 m row. Treatments included Sandea® (0.03 kg ha^{-1}) applied PRE, early POST (EPOST) at the snap bean 2nd trifoliate stage, and late POST (LPOST) at the 4th trifoliate stage, as well as Reflex® (0.35 kg ha^{-1}) applied EPOST and LPOST. A nontreated check was included for comparison. All plots, except for peas, were kept weed-free to isolate pea interference effects. Key findings include:

Pea Biomass Reduction: Sandea® PRE and EPOST significantly reduced mean per plot and per plant garden pea biomass (~55-60%) compared to the non-treated check (210.6 g $plot^{-1}$, 8.8 g $plant^{-1}$). Reflex® EPOST also reduced pea biomass (~35%), though not significantly. LPOST applications were less effective.

Snap Bean Yield: Mean marketable (USDA 1) and total snap bean yields were lowest with Sandea® LPOST (USDA 1 = 842 g m row^{-1}, total = 1136.1 g m row^{-1}). Marketable and total snap bean yields were equal to or higher than the non-treated check (USDA 1 = 1090.1 g m row^{-1}, 1362.1 g m row^{-1}) with Sandea® PRE, Sandea® EPOST, and Reflex® EPOST (USDA 1 = 1177.1 to 1283.4 g m row^{-1}, total = 1515.7 to 1625 g m row^{-1}). Reflex® applied LPOST caused less snap bean stunting relative to the other treatments, potentially mitigating yield loss.

Efficacy Timing: Both Sandea® and Reflex® suppressed pea growth rather than killing the plants, with early applications (PRE or EPOST, which corresponded to the V1-V3 pea stage) being more effective. These results highlight the importance of early intervention to reduce volunteer pea interference and potentially improve snap bean yields. However, more research is needed to assess broader impacts, including harvest efficiency, load quality, and the effects of sequential herbicide applications.

12

The Benefits of Overlapping Residual Herbicides in Weed Management Programs for Multiple Herbicide-Resistant Amaranthus Populations in Corn.

C. Austin[1], M. Kitt[2], D. Belles[2], T. Beckett

[1]*Syngenta Crop Protection, Palm, United States*, [2]*Syngenta Crop Protection, Greensboro, United States*

Multiple herbicide-resistant Waterhemp (*Amaranthus tuberculatus* [Moq.] Sauer.) and Palmer amaranth (*Amaranthus palmeri* S. Watson) have become difficult to control with traditional herbicide programs in corn. Field trials across multiple sites were implemented during 2024 to evaluate herbicide programs consisting of robust residual herbicide rates and overlapping residuals for the control of multiple herbicide-resistant *Amaranthus* spp. populations on LibertyLink® and Roundup Ready© 2. Corn. Data concluded that robust rates of overlapping residuals applied as a PRE fb early POST provided effective control of multiple herbicide-resistant *Amaranthus* spp. on LibertyLink® and Roundup Ready© 2. Corn.

Gap in pagination due to unavailable paper.

Page 13

14

Cornell University's Pigweed Identification Website and Guidebook

J. Petzoldt[1], A. Thomas[1], M. Bohrer[1], L. Sosnoskie[1]

[1]*Cornell AgriTech, Geneva, United States*

Pigweeds (*Amaranthus* spp.) are among the most common and noxious weeds encountered by US growers. Some species, such as Palmer amaranth (*A. palmeri*) can grow rapidly and outcompete crops for sunlight, water, and nutrients, achieving heights of up to 10 feet and producing up to one million seeds under optimal conditions. It can be difficult to distinguish pigweed species, which may co-occur within fields in the Northeast, without closely examining a combination of traits. Here we present an updated tabular and pictorial reference guide to pigweed species identification for nine common pigweed species including: Palmer amaranth, waterhemp (*A. tuberculatus*), redroot pigweed (*A. retroflexus*), Powell amaranth (*A. powellii*), smooth pigweed (*A. hybridus*), purple or livid amaranth (*A. blitum*), spiny amaranth (*A. spinosus*), prostrate pigweed (*A. blitoides*), and tumble pigweed (*A. albus*). Proper identification is critical, as seven of the nine species have evolved resistance to multiple herbicide modes of action. Furthermore, phenological knowledge, such as growth rate and form, can inform the timing and selection of weed management tools. Our goal is to facilitate timely and accurate pigweed identification towards an optimal pigweed management program. This information, including a pdf copy of the poster, can be found online at https://cals.cornell.edu/weed-science/weed-identification/pigweed-identification or by contacting Dr. Lynn Sosnoskie (lms438@cornell.edu).

15

Preliminary Evaluations of Postemergence Control Options for Alligatorweed and Hedge Bindweed in Nurseries and Landscapes

C. Harlow[1], J. Neal[1]

[1]*North Carolina State University, Raleigh, United States*

Perennial broadleaf weeds are challenging to manage in field nurseries and landscape plantings. Directed applications of glyphosate are generally effective on many weeds, but unsatisfactory control is often observed. Several postemergence herbicides were compared for control of alligatorweed (*Alternanthera philoxeroides*) and hedge bindweed (*Calystegia sepium*). The experiment was conducted in outdoor containers in 2022 and repeated in 2023. Hedge bindweed roots and alligatorweed stems were collected from local infestations in March of each year. Root pieces and stems with root initials were potted to 4-L containers using a pine bark substrate and top-dressed with slow-release fertilizer. Plants were grown for two months and pruned periodically to the edges of the containers to encourage branching and root sprouting. Herbicides were applied on July 6, 2022 and on June 28, 2023. Contact-action herbicides were reapplied at 2 weeks; systemic herbicides were reapplied at 4 weeks. Herbicides evaluated were 2,4-D amine, bentazon, clopyralid, flumioxazin + pyroxasulfone, glufosinate, glyphosate, rimsulfuron, tiafenacil, topramezone, triclopyr amine, and combinations of glufosinate with 2,4-D or topramezone. Each was applied at the manufacturer's recommended dose and with adjuvants as specified on the product label. Herbicides were applied with a CO_2-pressurized sprayer equipped with TeeJet 8008 flat fan nozzles and calibrated to deliver 50 GPA. Percent weed control was visually evaluated for 8 weeks. Glyphosate at 2 lb ai/A provided 100% control of alligatorweed but only 87% control of bindweed. Alligatorweed control with triclopyr amine @ 1.5 lb ai/A, 2,4-D amine @ 1.5 lb ai/A, or clopyralid @ 0.25 lb ai/A was not different from glyphosate. Flumioxazin + pyroxasulfone or rimsulfuron at 0.0625 lb ai/A suppressed alligatorweed 60 to 83%. Control with glufosinate was inconsistent, with regrowth occurring. Other treatments were ineffective on alligatorweed. Based on a Waller means separation procedure, the top-tier treatments for hedge bindweed control were: 0.71 lb ai/A flumioxazin + pyroxasulfone (100%), triclopyr (100%), 2,4-D (100%), rimsulfuron (94 to 95%), 0.36 lb ai/A flumioxazin + pyroxasulfone (91%), or glyphosate (87%). Glufosinate and tiafenacil controlled hedge bindweed as well as glyphosate but less than the best three treatments. Clopyralid and bentazon @ 32 oz/A each suppressed bindweed. Topramezone and the lower dose of bentazon each caused some temporary foliar injury but plants grew back. Tank mixing 2,4-D amine with topramezone or glufosinate did not improve overall control. The tank mix of 2,4-D + glufosinate provided less control compared to either product applied individually, suggesting potential antagonism that warrants further study. These data indicate that several herbicides can control alligatorweed and hedge bindweed as well or better than glyphosate. Results from these container trials need to be confirmed in natural, perennialized stands of these weeds.

Gap in pagination due to unavailable paper.

Page 16

17

Rutgers Japanese Stiltgrass Project: Empowering Extension Volunteers to Document Phenology Across New Jersey

J. Epiphan[1], M. Elmore[2], P. Nitzsche[1]

[1]Rutgers Cooperative Extension, Morristown, United States, [2]Rutgers University, New Brunswick, United States

Japanese stiltgrass (*Microstegium vimineum*) is an invasive weed species that has become problematic in lawns and detrimental to a variety of landscape types across New Jersey. Stiltgrass rapidly grows, seeds and spreads competing with desirable ornamental and native species. Control of stiltgrass requires an integrated approach of cultural and chemical means. Management techniques need to be timed for the most efficient and effective control. Yet, little is known about the temperature and timing required for stiltgrass germination and seed emergence. Rutgers Master Gardeners (RMG) and Rutgers Environmental Steward (RES) volunteers with known populations of Japanese stilgrass were recruited to participate in a project to document germination and seed emergence. Over a span of two years, 259 participants from 14 counties in New Jersey were trained through a webinar on the identification and biology of stiltgrass and the materials and methods of the project. A subset of attendees committed to monitoring their stiltgrass infestation site multiple times per week in Spring and late Summer in 2023 or 2024. 104 were able to identify stiltgrass and upload images of the seedlings germinating in diverse microclimates and recorded germination date. 48 volunteers recorded seed emergence data and uploaded images. Volunteers who mis-identified stiltgrass were informed to help increase their knowledge and skill in identifying stiltgrass and similar species. The temporal data of germination and seed emergence collected by the volunteers will be compiled and published to improve Japanese stiltgrass management recommendations. This project demonstrates the impact of facilitating volunteer data collection to inform control strategies of an invasive and detrimental weed species.

18
Spray drone, Flood Nozzle, and Ride-On-Sprayer Influences Smooth Crabgrass Response to Topramezone

S. Hale[1], N. Godara[1], J. Romero[1], S. Askew[1]

[1]*Virginia Tech, Blacksburg, United States*

Labor shortages in the physically demanding turfgrass management sector have spurred interest in alternatives to conventional hose-end sprayers, such as drones and ride-on sprayers, which promise benefits in terms of time management, efficiency, and accessibility. However, their efficacy in herbicide application for weed control in turfgrass systems has not been thoroughly examined. This research, conducted on July 9, 2024, at Virginia Tech in Blacksburg, VA, evaluates the performance of drones and ride-on sprayers against a traditional handheld spray gun with a flood nozzle in controlling smooth crabgrass (*Digitaria ischaemum*) using topramezone at rates of 18 or 36 g ae/ha on athletic-height (1.5 cm) Kentucky bluegrass turf. The turf was initially 25% covered by three- to four-tiller smooth crabgrass, which increased to over 90% in untreated areas by the study's end. Employing a randomized complete block design with four replications, the study featured a two (rates) by three (equipment) factorial treatment structure, including a non-treated check for baseline comparison. Application volumes were 28 L/ha for the drone, 234 L/ha for the ride-on sprayer, and 1121 L/ha for the spray gun. Results indicated a decline in smooth crabgrass control over time across all treatments, with peak efficacy observed at two and four weeks after treatment (WAT). A significant interaction between topramezone rate and application equipment was observed, where increasing the rate from 18 to 36 g/ha enhanced control by 20 to 40%, with the most substantial improvements seen with the spray gun. The ride-on sprayer achieved slightly better control than the spray gun, while the drone was approximately half as effective as ground-based equipment. This study contributes to the body of literature evaluating drone applications for weed control, showing variable outcomes. It also provides one of the few direct comparisons between ride-on sprayers and spray guns, suggesting that ride-on sprayers might offer superior weed control under similar conditions.

19

12 Years Investigating Cover Crop Diversity: Do Mixtures Improve Weed Suppression?

S. Hosler[1], J. Kaye[1], B. Baraibar Padró[2], C. Lowry[1]

[1]The Pennsylvania State University, University Park, United States, [2]Universitat de Lleida, Lleida, Spain

Farmers plant cover crops to provide a variety of ecosystem services, and organic producers rely on cover crops to partially or wholly replace the functions of synthetic chemical inputs, including herbicides. Previous studies have shown mixed results as to the efficacy of cover crops for providing weed suppression in organic grain fields. We investigated weed suppression as part of a long-term randomized block study on organic fields using eight common cover crop species planted as monocultures and in mixtures. Weeds and cover crops were sampled twice during the season: once in the fall before the first killing frost, and once in the spring immediately prior to cover crop termination. Sampling was conducted using quadrats, with aboveground biomass within the quadrats being harvested, then dried and weighed. We analyzed the effect of cover crop treatment (each species monoculture and each of several mixtures) on the amount of weed biomass as compared to a fallow treatment using ANOVA. Our results indicate that cover crop mixtures were more successful at repressing fall weeds than monocultures, with a trend toward mixtures with more species being more successful. However, patterns for spring weed suppression were much less clear. Our study provides a long-term, multi-treatment examination of cover crop strategies for weed suppression in organic grain production systems and indicates that cover crop mixtures can be a valuable tool for fall weed suppression.

Gap in pagination due to unavailable paper.

Page 20

21

The Root of Invasion: Cheatgrass (Bromus tectorum L.) Root System Architecture varies between genotypes collected in different habitats

K. Blocklove[1], N. Ali[1], J. Lasky[1], C. Lowry[1]

[1]*Pennsylvania State University, University Park, United States*

Understanding how invasive plants adapt to variable habitats can improve our ability to manage existing invasive populations and predict future invasions. Thus far, variability in aboveground traits have demonstrated the capacity for cheatgrass (*Bromus tectorum* L.), an invasive annual grass, to adapt to a wide range of habitat conditions. This project asks if root architecture shows similar trait variation, and if this variation associates with habitat climate. A global panel of genotypes was grown in a common garden in the Great Basin (Idaho, USA). Root-rhizosphere samples of three-leaf seedlings from 15 of these genotypes were collected, and root architecture was scanned. A Principal coordinates analysis was used to examine the associations between cheatgrass root architectural traits, as well as environmental conditions from the genotype site of origin. The preliminary analysis suggests that root architecture of genotypes from arid climates shows adaptations characteristic of drought adaptation. Future steps include amplicon barcoding of the rhizosphere microbiome associated with these samples and a greenhouse rhizobox collection of intact root-rhizosphere systems.

Gap in pagination due to unavailable paper.

Page 22

23

The Importance of Species Selection in Cover Crop Mixture Design

A. McKenzie-Gopsill[1], A. Mills[1]

[1]*Agriculture and Agri-Food Canada, Charlottetown, Canada*

Cover crops are increasingly being included in crop rotations as a mechanism to promote diversity and provide agroecosystem services including weed suppression. Recently, cover crop mixtures have increased in popularity in an attempt to provide a greater diversity in ecological services as compared to monocultures. Several recent studies, however, have failed to detect a positive effect of cover crop diversity on biomass production or weed suppression. Here we assessed biomass productivity and weed suppression in 19 cover crops seeded as monocultures and 19 mixtures of varying species composition and functional richness (2- and 3-species mixtures) of full-season cover crops in Atlantic Canada. Cover crop biomass production and weed suppression varied by species identity, functional diversity, and species richness. As cover crop biomass increased regardless of diversity, weed biomass declined. Highly productive forbs and grasses provided the greatest weed suppression in monoculture. In line with previous observations, mixtures were on average not more productive nor weed suppressive than the most productive monocultures. We observed that the inclusion of the highly productive species buckwheat and sorghum-sudangrass in a mixture increased stand evenness, productivity, weed suppression and spatiotemporal stability. Taken together our results suggest that effects of diversity on mixture productivity and weed suppression are species specific. This further demonstrates the importance of species selection in cover crop mixture design.

24
Tiafenacil Efficacy and Safety in Hop (Humulus lupulus L.) Production

L. Sosnoskie[1], M. Moretti[2]

[1]*Cornell AgriTech, Geneva, United States*, [2]*Oregon State University, Corvallis, United States*

The US is the largest hop (*Humulus lupulus* L.) producer in the world; in 2023, the ~22,000 cultivated ha were valued at $562 million. The majority of production (97%) occurs in Washington (74%), Idaho (16%), and Oregon (10%). Since the 2010s, hop cultivation has expanded outside the Pacific Northwest (PNW), particularly in the Great Lakes region and New York State (NYS), driven by brewers seeking locally sourced ingredients.

Weeds compete with hops for resources and disrupt practices like training and harvest. They can also harbor spider mites and alter the microclimate, increasing the risk of downy mildew. Currently, carfentrazone is used for controlling broadleaf weeds and managing basal canopy foliage to suppress pests and pathogens. In 2022–2023, IR-4 Project trials in NYS evaluated tiafenacil (Reviton®), a Group 14 postemergence herbicide effective against broadleaf and small annual grasses, in eastern hop production.

The trial, conducted at Cornell AgriTech, was planted with four varieties (‘Cascade,’ ‘Centennial,’ ‘Alpha Aroma,’ and ‘Nugget’) on 5.5 m trellises. Canada thistle and various grasses dominated the site. Herbicide treatments included: 1) nontreated check (NTC), 2) Aim ® (FMC) 0.03 kg ai ha^{-1} + COC 1% v/v at ABCD timings; 3) Reviton® 2.83 SC (Helm) 0.05 kg ai ha^{-1} + MSO 1% v/v at ACD timings, 4) Reviton® 2.83 SC 1.0 kg ai ha^{-1} + MSO 1% v/v at ACD timings, 5) Reviton® 2.83 SC 0.05 kg ai ha^{-1} + MSO 1% v/v at BCD timings, 6) Reviton® 2.83 SC 1.0 kg ai ha^{-1} + MSO 1% v/v at BCD timings. Timings included applications during winter dormancy (A), early spring shoot emergence (B), and post-directed sprays at crop heights of 1.8-2.4 m (C) and 3.7-4.9 m (D).

Reviton® consistently outperformed Aim® in reducing weed cover, with higher efficacy at the 1.0 kg ha^{-1} rate. Averaged over observation dates, weed cover in the Aim® plots was 41% while weed cover in the Reviton® plots was 23%. Reviton® was more effective at managing Canada thistle; Canada thistle control in the Aim® plots averaged 41% while control in the Reviton® plots averaged 64%. Grasses were only suppressed, not eradicated, by Reviton®. Emerging hop shoots treated at timing B exhibited more necrosis in Reviton® treatments compared to Aim®. This burn-back effect is desirable for synchronizing shoot growth and boosting yields. Post-directed applications also caused greater basal foliage necrosis in Reviton® treatments, aiding pathogen management. Importantly, no injury was observed on climbing bines or foliage above the treated zone. Herbicide treatments did not impact plant height or biomass, with cultivar differences driving growth and yield variability. ‘Alpha Aroma’ and ‘Cascade’ were more vigorous than ‘Centennial’ and ‘Nugget.’ Results in 2023 aligned with 2022 trials and PNW findings, confirming that Reviton® effectively reduces weed pressure and facilitates hop foliage management in eastern hop production.

25

Evaluating the Effect of Containerized Sweetpotato (Ipomoea batatas) on Herbicide Tolerance

S. Crawford[1], K. Jennings[1], D. Monks[1], D. Jordan[1], J. Schultheis[1], A. Gorny[1], L. Moore[1], K. Goble[1], H. Thome[1]

[1]*North Carolina State University, Raleigh, United States*

Sweetpotato (*Ipomoea batatas*) is traditionally established by transplanting nonrooted cuttings directly into the field. Limited research has been conducted on planting sweetpotato transplants grown in the greenhouse. This method allows the cuttings to form roots and be more established before being transplanted into the field which has the potential to increase herbicide tolerance. Therefore, studies were conducted at the Horticultural Crops Research Station in Clinton NC to evaluate the effect of containerized sweetpotato plug plants on herbicide tolerance. Containerized transplants were grown in the greenhouse for two wk prior to transplanting. A traditional nonrooted sweetpotato cutting treatment was included for comparison. Herbicide treatments included fluridone at 336 g ai ha^{-1}, flumioxazin at 107 g ha^{-1} plus sulfentrazone at 210 g ai ha^{-1}, flumioxazin PRE at 107 g ha^{-1} followed by (fb) pyroxasulfone at 292 g ai ha^{-1}, flumioxazin PRE at 107 g ai ha^{-1} fb linuron at 420 g ai ha^{-1}, and flumioxazin PRE at 107 g ai ha^{-1} fb metribuzin at 315 g ai ha^{-1}. A nontreated check was included for comparison. Vegetative propagation method did not have an effect on herbicide injury, but containerized transplants did reduce total and marketable yield. Flumioxazin fb metribuzin and sulfentrazone plus flumioxazin decreased total yield while fluridone, flumioxazin fb linuron, and flumioxazin fb pyroxasulfone did not affect total yield. Increased culls were observed from the containerized sweetpotato transplant treatment. Inversely, no. 1 storage roots was greater for traditionally planted sweetpotato.

26
Combine Performance Metrics While Using Seed Impact Mills.

M. Flessner[1], E. Russell[1], M. Bagavathiannan[2], K. Bejleri[3], S. Chu[2], W. Crane[3], E. Law[4], S. Mirsky[3], D. Sarangi[5], L. Shergill[6], M. Vangessel[7]

[1]Virginia Tech, Blacksburg, United States, [2]Texas A&M University, College Station, United States, [3]USDA ARS, Beltsville, United States, [4]The Ohio State University, Columbus, United States, [5]University of Minnesota, St. Paul, United States, [6]Colorado State University, Ft. Colins, United States, [7]University of Delaware, Georgetown, United States

Seed impact mills are used to implement harvest weed seed control (HWSC) by killing weeds seeds as they exit the combine, reducing inputs to the weed seed bank. Seed impact mills, such as the Redekop Seed Control Unit (SCU) and the Harrington Seed Destructor (iHSD) draw power from the combine, which can influence fuel consumption, travel speed, and horsepower demand. Farmers are keen to know these metrics to know the true cost of operating a seed impact mill. Therefore, the objective of this research was to determine and compare fuel use, travel speed, and engine capacity use for the SCU and iHSD across crops, combines, and harvest conditions. Seed impact mills (SCU (N=4) and iHSD (N=3)) were installed on cooperating farmers' Case IH (N=3) and John Deere (N=4) combines. Combine metrics, crop yield, and grain moisture (as reported on the combine's display) were recorded during harvest of 300 ft long passes in rice, soybean, corn, and wheat in 6 states from 2022 to 2024. Data were collected while the seed impact mill was both engaged (i.e. performing HWSC) and disengaged (i.e. performing conventional harvest) on the same combine within site-year. Chaff moisture was also collected manually from samples exiting the seed impact mill during use. Data were analyzed using ANOVA and subsequent means separation using Fisher's Protected $LSD_{(0.1)}$ by crop across harvest conditions and combine configurations. A model of HWSC treatment, seed impact mill, and their interaction was used to evaluate fuel use, travel speed, and engine capacity. To compare the iHSD and SCU, relative differences in fuel use, travel speed, and engine capacity were calculated by subtracting the observed value when the mill was engaged from the value when it was disengaged. No interactions were detected between seed impact mill brand and HWSC treatment. Operating seed impact mills increased fuel use by 2.5 and 2.8 gal/hr and engine capacity consumed by 11.1 and 8.8% for both iHSD and SCU, respectively, across crops. Travel speed was reduced using the iHSD by 0.35 MPH across crops but not reduced with the SCU. After accounting for the combine without the seed impact mill, travel speed during soybean harvest was the only difference detected when comparing the iHSD and SCU. The iHSD decreased speed by 0.64 MPH compared to a 0.02 MPH increase with the SCU. Combine operators should expect increased fuel use and potentially decreased travel speed when operating seed impact mills in corn, rice, wheat, and soybean, and thus increased harvest costs and time.

27
Finding the Fix: Root Systems Architecture Variation and Nitrogen Fixation.

N. Ali[1], K. Blocklove[1], C. Lowry[1], J. Lasky[1]

[1]*The Pennsylvania State University, University Park, United States*

Cheatgrass (*Bromus tectorum L.*) is an invasive annual grass that is associated with greater bioavailability of soil nitrogen (N). Plant interactions with the soil microbiota affect resource cycling functions, including mechanisms which input N: fixation or saprotrophism. This project asks a) if cheatgrass shows genotypic variation in the relative enrichment of isotopic ^{15}N, a proxy for microbial fixation. We also explore the effect of competition with crested wheatgrass (*Agropyron cristatum*), the most broadly successful restoration grass. To do this, we performed a full-factorial randomized complete block experiment with rhizoboxes of 15 genotypes with intraspecific or interspecific (versus crested wheatgrass) competition. The homogenized shoot tissue was combusted to quantify isotopic nitrogen ($\Delta\,^{15}$N), which associates with the relative contributions of N fixation. Preliminary results indicate that genotype, but not competitor, affects Δ^{15}N. Next steps will test for associations with adaptive life history strategies or habitat conditions of the original genotype. Additional repetitions of the experiment are ongoing. Insights into the ^{15}N content of cheatgrass shoot tissue can inform the prediction of invasions and support decision making in land management.

28

Monitoring the distribution of herbicide-resistant Amaranthus spp. in Pennsylvania

I. Moisinho[1], C. Brunharo[1]

[1]The Pennsylvania State University, University Park, United States

The *Amaranthus* genus comprises summer annual weeds that are problematic in agricultural and horticultural settings. Due to the increased reliance on herbicides for weed management, herbicide resistance has emerged as a significant concern. In the United States, over 170 cases of herbicide resistance have been documented, including three cases in Pennsylvania involving resistance to PSII, ALS, and EPSPs inhibitors in corn, soybean, and tomato fields. This study aimed to assess the distribution and evolution of herbicide-resistant *Amaranthus* species in Pennsylvania. Populations were collected from fields where weed control was inadequate at the end of the growing season. Mature seeds from individual plants within each field were pooled and screened with post-emergent herbicides. Seedlings (2-4 leaves) were sprayed with glyphosate (1,107 g ai/ha), dicamba (1,127 g ai/ha), chlorimuron (26 g ai/ha), 2,4-D (961 g ai/ha), mesotrione (211 g ai/ha), fomesafen (507 g ai/ha), atrazine (1,348 g ai/ha), imazethapyr (105 g ai/ha), and glufosinate (639 g ai/ha). At 21 days after application (DAA), plant injury was assessed and plotted on maps to visualize the distribution of resistance. Sampled fields included corn, soybean, teff, orchards (peach, apple), beans, strawberry, and non-defined (ND) areas. A total of 36 *Amaranthus* populations were collected, inclusing *A. palmeri, A. retroflexus, A. tuberculatus, and A. hybridus*. The highest survival rates (injury < 50%) were observed with the ALS inhibitors imazethapyr (56%) and chlorimuron (22%), followed by the PSII inhibitor atrazine (22%) and the EPSPs inhibitor glyphosate (22%). We did not identify resistant population to fomesafen, 2,4-D and glufosinate. A total of 13, 8, 7 and 1 population were resistant to 1, 2, 3 and 4 mechanism of action, respectively. Lower survival rates were observed with the HPPD inhibitor mesotrione (6.5%) and the auxin mimic dicamba (2.2%). It is important to identify herbicide resistance to plan weed management strategies and avoiding spread to non-infested areas. Dose-response experiments will be conducted to determine the resistance index of the populations, and molecular markers will be developed for rapid detection of resistance.

29

Response of Seeded Hemp to Pre- and Post-Emergence Herbicides

L. Sosnoskie[1], T. Besançon[2]

[1]*Cornell AgriTech, Geneva, United States*, [2]*Rutgers University, New Brunswick, United States*

Sonalan® HFP is the only herbicide currently registered for hemp, making its effective use critical for crop establishment. At Cornell AgriTech, a 2024 trial evaluated its safety when applied pre-emergence (PRE) versus pre-plant incorporated (PPI) in 'Białobrzeskie' hemp. Sonalan® HFP was applied at 1.05 kg ai ha^{-1}, with PPI incorporated to a depth of at least two inches before planting and PRE applied after planting. Weed and hemp counts were assessed weekly to bi-weekly. Hemp and broadleaf weed densities and biomass estimates were unaffected by application treatment, Grass density was 43-83% lower with PPI than PRE. Under exceptionally dry conditions, grass biomass was significantly higher in PRE-treated areas, suggesting that under dry conditions, as was seen in June and July 2024, PPI may better for suppressing annual grasses, which are a target of Sonalan®.

In a 2022 Cornell study, the safety and performance of PRE-applied herbicides across a variety of WSSA groups (2, 13, 14, 15) were tested on 'Anka' hemp. Herbicides were applied PRE immediately after planting. All treatments except Nortron® suppressed weeds effectively; all tested herbicides caused significant and unacceptable early-season crop injury, with stunting reaching up to 100%. Even the least injurious herbicides, such as Dual Magnum®, Nortron®, and Sandea®, reduced crop emergence by over 50%.

A 2024 trial at Rutgers Agricultural Research and Extension Center in New Jersey examined PRE herbicides in 'Białobrzeskie' hemp. Treatments included Sonalan® at 1.26 kg ai ha^{-1}, Linex® at 0.70 kg ai ha^{-1}, Linex® at 1.28 kg ai ha^{-1}, Outlook® at 0.74 kg ai ha^{-1}, Outlook® at 1.48 kg ai ha^{-1}, Zidua® at 0.13 ks ai ha^{-1}, and Zidua® at 0.26 kg ai ha^{-1}. Mean weed cover was greatest in the untreated check rising from 39% at 7 DAT to 89% at 28 DAT. Sonalan®HFP at 1.26 kg ai ha^{-1}, Linex® at 0.70 and 1.28 kg ai ha^{-1}, and Zidua® at 0.13 ks ai ha^{-1} and 0.26 kg ai ha^{-1} provided the most effective weed control; mean weed cover across these treatments ranged from 30% to 49% at 28 DAT. Mean weed cover in the low and high Outlook® treatments was 65% and 70% at 28 DAT, respectively. In addition to providing some of the most effective weed control, Sonalan® HFP at 1.26 kg ai ha^{-1} and Linex® at 0.70 and 1.28 kg ai ha^{-1} were also the least injurious herbicides evaluated in the trial. Sonalan®HFP at 1.26 kg ai ha^{-1} increased mean hemp biomass (480%) and plant density (325%), relative to the untreated check; density improvements were numerical and not statistical. The Linex® treatments also increased mean hemp biomass and density (258% to 673%) compared to the check plots, although density gains were not significant. Outlook® and Zidua® did not improve upon hemp biomass or stand counts.

Sonalan®HFP performance may be improved by PPI use under adverse weather conditions. Also, these results align with earlier findings by highlighting hemp's sensitivity to many soil-applied herbicides. Linex® is a promising chemistry that requires additional evaluation in hemp.

30

Weed Community Composition in Agrivoltaic Cropping Systems.

D. Russell[1], A. Ditommaso[1], C. Marschner[1]

[1]*Cornell University, Ithaca, United States*

Agrivoltaic (AV) systems have the potential to increase land-use efficiency through co-location of solar energy and agricultural production. These systems are becoming increasingly important as demand for solar energy and food production rises, particularly because solar installations are frequently sited on agricultural land. While AV systems have been shown to influence abiotic conditions, limited research has been conducted in the Northeastern United States, and no published studies have examined the effects of these systems on weed pressure. This study aimed to assess the impact of an AV cropping system on weed community diversity and biomass. We hypothesized that 1.) weed diversity and biomass would be lower in the AV treatment group, and 2.) that average air temperature and photosynthetic photon flux density (PPFD) would be lower and soil moisture would be higher in the AV treatment group. Four replicate plots of strawberry (*Fragaria x ananassa*), raspberry (*Rubus idaeus*), radish (*Raphanus sativus*), radicchio (*Cichorium intybus*), lettuce (Lactuca sativa), and spinach (*Spinacia oleracea*) were planted in control and AV treatment groups. AV treatment plots were planted directly between panel rows and control plots were planted on an adjacent unshaded plot of land. Sensors recorded air temperature, soil moisture, and PPFD in both treatments. Crops were planted on September 5-13, 2024. Weed species counts were recorded in weedy check quadrats at three times during the fall growing season, and weed biomass was collected at the final sampling to determine dry biomass. Shannon Diversity Index did not differ between sampling dates or treatments. Weed dry biomass was higher in AV (mean = 1.86 g) than in the control (mean = 0.10 g, $p < 0.01$). There was a ~35% reduction in direct sunlight within the array, but soil moisture and air temperature did not vary between treatments. These results indicate that AV systems may be able to reduce weed pressure without compromising ecosystem serviced provided by weed communities, as shown by reduced biomass in the AV treatment and similarity of Shannon Diversity Index over time between AV and control treatments. Environmental data suggest that lower light availability in the AV treatment is likely responsible for reduced weed biomass. Future research will focus on measuring weed pressure throughout the summer growing season when weed pressure is likely to be higher. Long-term studies are needed to evaluate the impact of agrivoltaic systems on composition of the weed community and weed seedbank over time.

31
Determining Pre-Emergent Herbicide Options for North Carolina Pollinator Species.

H. Lee[1], C. Cahoon[1], Z. Taylor[1], B. Dean[1], J. DeSanctis[1], J. Forehand[1]

[1]*North Carolina State University, Raleigh, United States*

During the summer of 2024, a study was conducted to evaluate wildflower species tolerance to various herbicides applied PRE-transplant and POST-transplant. The wildflower species evaluated were black-eyed Susan *(Rudbeckia hirta)*, purple coneflower *(Echinacea purpurea)*, Scarlet bee-balm *(Monarda didyma*), and Shasta daisy *(Leucanthemum superbum)*. These species were transplanted and treated on May 30, 2024, at the Central Crops Research Station in Clayton, NC. Treatments were replicated 4 times with a randomized strip plot design. The following herbicides were applied PRE-T (rate in g ai ha^{-1}): pyrithiobac (76), mesotrione (140), fluometuron (1212), sulfentrazone (140), indaziflam (58), a premix of pyroxasulfone and flumioxazin (46 + 34), and fluridone (168). The following herbicides were applied POST-T: pendimethalin (1064), pendimethalin (2128), S-metolachlor (1092), and pendimethalin + *S*-metolachlor (2156). An untreated check was included for comparison. When required by label, crop oil concentrate, methylated seed oil, and non-ionic surfactant were added at a one-half percent, one percent, and one-quarter percent volume-per-volume basis, respectively. Plots were 3 m wide by 9 m long with two treated transplant rows. Treatments were applied using a CO_2 backpack sprayer equipped with Teejet flat-fan AIXR 11002 nozzles, calibrated to deliver 140 L $ha-^{1}$. Wildflower injury was visually estimated one, two, four, and eight weeks after treatment, and plot density was recorded 8 weeks after treatment. When pendimethalin and *S*-metolachlor were applied POST-T, minimal injury and reductions in plant density were observed across all species. For black-eye Susan, acceptable (0-15% injury) PRE-T treatments were indaziflam, sulfentrazone, pyrithiobac, fluridone, and mesotrione. Those herbicides caused little to no injury and minimal reductions in plant density. For Scarlet bee-balm, indaziflam, sulfentrazone, and mesotrione caused little to no injury and minimal reductions in plant density compared to the untreated check. For purple coneflower, indaziflam, sulfentrazone, pyrithiobac, fluridone, mesotrione, and fluometuron caused little to no injury and minimal plot density reduction compared to the untreated check. For Shasta daisy, indaziflam, sulfentrazone, fluridone, mesotrione, and fluometuron caused little to no injury and minimal plot density reduction compared to the untreated check. In conclusion, this data suggests that several herbicide options have the potential to be safely integrated over the top of these wildflower species. Herbicides that appear safe will be recommended for further replicated testing. After further testing, 24(c) labels can be perused so that these products may be incorporated into weed management plans by the NC DOT wildflower program.

32

Impact of Plant Height on the Effectiveness of Atrazine, Dicamba, Glufosinate, Glyphosate, Lactofen, and 2,4-D on New York Common Lambsquarters (Chenopodium album) Populations.

E. Nelson[1], V. Kumar[1]

[1]*Cornell University, School of Integrative Plant Science, Soil and Crop Sciences Section, Ithaca, United States*

Common lambsquarter (*Chenopodium album*) is one of the most troublesome weed species in agronomic crops in the United States, including New York. Uncontrolled infestation of common lambsquarters is known to cause considerable yield losses in corn and soybeans. Growers primarily rely on POST-emergence herbicides for control of common lambsquarters and other weed species in corn and soybeans. However, application timing of POST herbicides is crucial for improved effectiveness. The objective of this research was to understand the impact of plant height on the effectiveness of POST herbicides for control of common lambsquarters. Greenhouse experiments were conducted in fall 2024 at Cornell University Bioclimatic Laboratory in Ithaca, NY. Two common lambsquarter populations (Ferguson and Knight) collected from two separate corn fields were investigated at two different plant heights (10 cm and 20 to 30 cm). Six commonly used POST herbicides, including atrazine, glyphosate, 2,4-D, dicamba, glufosinate, and lactofen were tested. A nontreated weedy check was also included for treatment comparison. All treatments were arranged in randomized complete block design in factorial combinations (two populations x two heights x six herbicides) with 15 replications. Selected POST herbicides were applied in cabinet spray chamber using AIXR nozzles at 141 L ha^{-1} spray solution. Data on percent visual control were collected at 14, 21, and 28 days after application (DAA). Shoot biomass was collected at 28 DAA and weighed to estimate percent biomass reduction. Data was subjected to analysis of variance using PROC MIXED in SAS Studio to determine differences in herbicide performance between growth stages and populations. Results showed that 2,4-D, atrazine, and dicamba provided the highest control rates (> 90%) across populations, regardless of plant height at 28 DAA. Herbicides with contact mode of action, such as lactofen and glufosinate had lower effectiveness on both populations when plants were 20-30 cm. These findings highlight the importance of early applications of POST herbicides (especially lactofen, glyphosate, and glufosinate). The study also revealed relatively lower effectiveness of glyphosate on Ferguson compared to the Knight population. This study also concludes that if late applications are unavoidable (plant heights > 20 cm), POST applied atrazine, dicamba, and 2,4-D are effective options to control common lambsquarter.

33
Comparison of Black Oats and Cereal Rye as Cover Crops for Weed Suppression.

J. Beville[1], M. Flessner[1]

[1]*Virginia Tech, Blacksburg, United States*

Cover crops play an important part in various cropping systems and have multiple agronomic and environmental benefits. Farmers in the mid-Atlantic region of the United States favor cereal rye (*Secale cereale* L.) due to its copious biomass production allowing for weed suppression. However, cereal rye also has a high carbon to nitrogen (C:N) ratio, which can result in nitrogen immobilization. Because of this, some Virginia farmers have looked towards using a grass cover crop that may have a lower C:N ratio: black oats (*Avena strigosa* L.). Experiments in two Virginia locations were conducted to compare biomass production and composition and the overall weed suppression abilities of 'Late Abruzzi' cereal rye and 'Cosaque' black oats. Biomass samples were collected from each plot by hand using a random 0.25 m^2 quadrat throughout the growing season until termination in early May. Weed density data on the five most abundant weed species within each location were also collected using two 0.25 m^2 quadrats randomly placed within the middle interrow of each plot. Weed density counts occurred one month after soybean (*Glycine max* L.) planting, just prior to postemergence herbicide application. Cover crop biomass data indicated that there were no statistical differences at either location at each collection timing. Lignin content was similar across locations. Over time, lignin content increased for cereal rye but decreased for black oats resulting in a significant difference at planting of 5.0% compared to 2.1%, respectively. For C:N ratio, there were no differences across species or location. Over time, the C:N ratio increased from 22.2 four weeks before planting to 33.4 at planting. Total weed density data from both locations demonstrated that both cover crops resulted in similar weed suppression. Therefore, the results indicate that not only can farmers potentially produce adequate amounts of biomass with black oats, but can also achieve similar weed suppression results when compared to the industry standard, cereal rye.

34
Control of Glyphosate-resistant Palmer Amaranth Populations from New York and Connecticut.

M.. Rieley[1], V.. Kumar[1], M.. Stanyard[2], J.. Aulakh[3], P.. Bajwa[1], H.. Scatena[1]

[1]Cornell University, School of Integrative Plant Science, Soil and Crop Sciences Section, Ithaca, United States, [2]Cornell Cooperative Extension, Ithaca, United States, [3]The Connecticut Agricultural Experiment Station, Windsor, United States

Palmer amaranth (*Amaranthus palmeri*) is one of the most problematic weed species in the United States, with documented resistance to eight distinct herbicide classes. Season-long infestations of Palmer amaranth can result in yield losses up to 91% in corn, 79% in soybean, and 78% in cotton. Recent reports have confirmed glyphosate-resistant (GR) Palmer amaranth populations in New York (NY) and Connecticut (CT), highlighting the urgent need for alternative control measures. The objective of this research was to evaluate the effectiveness of alternative post-emergence (POST) herbicides for managing GR Palmer amaranth populations. Greenhouse experiments were conducted in 2024 at Cornell University's Guterman Bioclimatic Laboratory in Ithaca, NY. Three confirmed GR Palmer amaranth populations from NY, CT, and KS were investigated. Thirteen POST herbicide treatments, including 2,4-D choline, dicamba, glyphosate, glufosinate, lactofen, and saflufenacil alone or in combinations were evaluated. A nontreated check was also included for treatment comparison. Plants were grown in 10-cm diameter plastic pots containing potting mix, and herbicides were applied when plants reached 7.5 cm height using a cabinet spray chamber equipped with AIXR nozzles delivering 140 L ha^{-1} spray solution. Visual control ratings were assessed at 14, 21, and 28 days after treatment (DAT), and shoot biomass reduction was measured at 28 DAT. Data were subjected to analysis of variance using PROC MIXED in SAS, and means were separated using Fisher's protected LSD ($P < 0.05$). Results showed that glyphosate provided only 8%-18% control across all three populations, confirming glyphosate resistance. Alternative POST herbicides, including 2,4-D, dicamba, glufosinate, lactofen, saflufenacil and various combinations provided excellent control (94-100%) of GR Palmer amaranth from CT and KS populations. The NY population showed similar high control levels with most treatments, except for lactofen and saflufenacil + 2,4-D combinations. Consistent with visual control ratings, most alternative POST herbicides reduced shoot biomass by ≥90% across all populations, while glyphosate treatments resulted in minimal biomass reduction (8-66%). These results conclude that alternative POST herbicides containing multiple modes of action can effectively control GR Palmer amaranth populations from NY, CT, and KS, providing effective management options for growers.

35

Pyridate Admixtures Improve Weed Control.

J. Romero[1], N. Godara[1], S. Hale[1], S. Askew[1], B. Corbett[1]

[1]*Virginia Tech, Blacksburg, United States*

Pyridate is a contact herbicide in the photosystem II (PSII) inhibitor group. It was discovered in 1976, discontinued in 2007 and recently reintroduced to the market. Pyridate has proved faster and enhanced weed control when mixed with other herbicides. Previous studies have demonstrated enhanced weed control of tank-mixing herbicides with pyridate when used in crop production. However, no studies have been conducted to asses effectiveness in turfgrass systems. This study aimed to determine if combining pyridate with Sublime™ 494 g ai ha^{-1} improves broadleaf weed control compared to Sublime™ alone at different rating times. The study also compare this combination against industry-standard herbicides for control of multiple broadleaf weed species at one month after treatment. Twelve field trials were conducted from 2021 to 2024 at the Glade Research Center in Blacksburg, VA. Using a randomized complete block design. Treatments included Sublime™ 494 g ai ha^{-1}, Sublime™ 494 g ai ha^{-1} + pyridate 350 g ai ha^{-1}, and established industry standards (TZone, Sure Power, Cool Power, Escalade II). Weed control and turfgrass phytotoxicity were evaluated at 7, 14, 28, 56, and 84 days after treatment (DAT), data collected included digital pictures, drone imaging, and visual assessments. Results showed that tank-mixing pyridate with Sublime™ consistently accelerated weed suppression and sustained it longer compared to Sublime™ alone. Furthermore, the performance of Sublime™ 494 g ai ha^{-1} + pyridate 350 g ai $^{-1}$ was comparable to that of industry-standard herbicides.

36

Effect of green pigment and optical brightener on UV floral reflectance and pollinator foraging of lawn weeds

N. Godara[1], J. Romero[1], D. Koo[1], S. Askew[1]

[1]*Virginia Tech, Blacksburg, United States*

Pollinators risk exposure to insecticide residue when visiting weedy flowers in urban landscapes. Previous researchers documented that ultraviolet (UV) reflectance of floral features affects pollinator foraging behavior. We hypothesized that turf protection products (TPPs) would alter the UV reflectance of weedy flowers and associated pollinator foraging, altering a potential method to deter them from visiting treated flowers. A study was conducted at Blacksburg, VA in 2023 to assess the effect of TPPs on the UV-A (315-400 nm) reflectance at 4 hours after treatment and morphology of dandelion flowers and associated honeybee visitation. Treatments included a nontreated control, TurfScreen™ (ZnO, TiO_2, phthalocyanine), Exteris SG® (fluopyram, trifloxystrobin, phthalocyanine); Tinopal® (benzenesulfonic acid); Coppertone® Sunscreen (avobenzone, homosalate, octisalate, oxybenzone); and Coppertone® Sunscreen (sprayed with atomizer). Nontreated dandelion flowers reflected 18% UV-A at the petal apex, but UV-A reflectance was reduced to 6% by the residue of Tinopal and Sunscreen. Tinopal and Sunscreen treatments also reduced honeybee foraging visits by 50% for up to 2 d after treatment. Similarly, Exteris SG also reduced UV-A reflectance to <12%, but this reduction was caused by the morphological alteration in flowers and not residue. Honeybee foraging on dandelion flowers was transiently suppressed after Exteris SG, Tinopal, and Sunscreen treatment but recovered at 14 d after herbicide treatment. TPPs, like the optical brightener Tinopal, could be used in conjunction with insecticide sprays to reduce honeybee foraging in treated areas and potentially reduce bee exposure to harmful insecticides. Future research will investigate insecticide formulations comprised of compounds like benzenesulfonic acid that alter UV signals and transiently deter pollinators following treatment.

37

Effect of Glyphosate Formulation, Rate, and Application Timing on Grass Control

B. Dean[1], C. Cahoon[1], Z. Taylor[1], J. de Sanctis[1], H. Lee[1]

[1]*North Carolina State University, Raleigh, United States*

In recent years, numerous complaints have surfaced regarding Texas millet (*Urochloa texana* R. Webster) surviving multiple applications of glyphosate. Many complaints concern Roundup PowerMAX® 3, often applied alone to control late-season escapes. To address these concerns, a field study was conducted in 2024 near Rocky Mount, NC, to determine if glyphosate formulation, application rate, and Texas millet (*Urochloa texana* R. Webster) size affect control. The treatment structure was a 3×3×2 factorial arranged in a randomized complete block design with 4 replications. Treatments factors included three glyphosate formulations, three application rates, and two Texas millet sizes, plus a check. The three glyphosate formulations were Roundup PowerMAX® 3 (potassium salt; 575 g ae/L), Makaze® (isopropylamine salt; 356 g ae/L), and Durango® (dimethylamine salt; 479 g ae/L). Each herbicide was applied at 840, 1,260, and 1,681 g ae/ha to 15 and 30 cm tall Texas millet. Applications were made using a CO_2-pressurized backpack sprayer calibrated to deliver 140 L ha^{-1} at 207 kPa, equipped with AIXR 11002 flat-fan nozzles. Weed control was visually estimated 7, 14, and 21 days after treatment. At 14 DAT, Durango® and Makaze® were more effective than Roundup PowerMAX® 3 (80%), resulting in 85 and 84% control, respectively. By 21 DAT, no differences in control were observed between glyphosate formulations, with each achieving ≥ 90% control. Regardless of the herbicide or application rate, Texas millet was 100% controlled when treated at 15 cm in height. As expected, glyphosate applied at 1,681 g ae ha^{-1} recorded the greatest control of 30 cm tall Texas millet (90%), whereas 840 and 1,260 g ae ha^{-1} rates accounted for 74 and 81%, respectively. Overall, this study suggests that each glyphosate formulation can effectively control Texas millet that is 15 cm tall; however, a rate of 1,681g ae ha^{-1} may be needed to control 30 cm tall Texas millet in late-season.

38

Confirmation of a multiple resistant goosegrass (Eleusine indica) population in North Carolina.

R. Argueta[1], D. Contreras[1], C. Bradshaw[1], J. Alsdorf[1], C. Cahoon[1], W. Everman[1]

[1]*North Carolina State University, Raleigh, United States*

In 2022, near Winfall, NC, a farmer indicated that the goosegrass population survived glyphosate and clethodim applications in a cotton field. The objective was to evaluate the putative-resistant population's response at different glyphosate, clethodim, and fluazifop rates. Two susceptible populations from Wake County (NC1 and NC3) were used to determine differential susceptibility. Glyphosate was applied at 1.34 kg ae ha^{-1} (1X), clethodim at 102 g ai ha^{-1} (1X), and fluazifop at 140 g ai ha^{-1} (1X). Goosegrass control was evaluated at 7, 14, and 21 days after treatment (DAT), and mortality and biomass were collected at 21 DAT. Results were analyzed in SAS 9.4 and subsequently graphed in Sigmaplot using a 4-parameter logistic equation. The resulting resistance ratio (R/S) of 9.81 and 16.06 suggests that the Winfall population is glyphosate-resistant. The resulting R/S 2.58 to 3.25 indicates that the Winfall population may be clethodim-resistant. The resulting R/S 12.41 to 13.26 suggest that the Winfall population is fluazifop-resistant. The differential susceptibility to glyphosate, clethodim, and fluazifop exhibited by the putative-resistant population compared to the NC1 and NC3 populations suggests that the population is two-way resistant. These results provide evidence that the Winfall population goosegrass has evolved resistance to glyphosate, clethodim, and fluazifop.

39

Influence of Cover Crop and Sensitivity Setting on See & Spray™ Ultimate vs See & Spray Premium

D. Contreras[1], B. Pendleton[1], C. Bradshaw[1], J. Alsdorf[1], R. Argueta[1], C. Cahoon[1], L. Lazaro[2], W. Everman[3]

[1]North Carolina State University, Raleigh, United States, [2]Blue River Technology, Santa Clara, United States, [3]Iowa State Univeristy, Ames, United States

See & SprayTM (S&S) technology is a precision agriculture sprayer system that utilizes computer vision and machine learning to target weeds. Among its commercial platforms, both S&S Premium and S&S Ultimate provide the ability to target weeds within crops. However, a key difference between the two platforms is that S&S Premium features a single tank, while S&S Ultimate has dual tanks, allowing for simultaneous broadcast residual application alongside targeted spray application. S&S operators can adjust the sprayer sensitivity setting depending on their goals; a lower sensitivity setting requires higher confidence to identify a weed, whereas confidence increases with weed size, while higher sensitivity settings reduce the confidence needed for spraying. Cover crops are commonly used as a weed suppression tactic. Integrating multiple weed management strategies can enhance weed control, therefore a study was conducted with the objective of assessing the efficacy of See & Spray Premium and See & Spray Ultimate as influenced by cover crop and sprayer sensitivity settings. The experiment treatments followed a factorial design with Factor A: cover crop vs. no cover crop, Factor B: low, medium, or high sensitivity, and Factor C: S&S Premium or S&S Ultimate platform, along with two untreated treatments (with and without a cover crop) for comparison. Herbicide applications were made with an Agronomy Test Machine (ATM), a condensed version of the S&S Ultimate platform adapted for small plot research, at preemergence, early postemergence and mid-postemergence timings. Weed control data were collected at the time of each postemergence application and 14 days after mid-postemergence. Cover crop provided significant suppression of Palmer amaranth (*Amaranthus palmeri* S.) and goosegrass (*Eleusine indica* G.) at all rating dates. Sprayer sensitivity and S&S platform did not significantly impact weed control. This study demonstrated that the efficacy of See & Spray Premium and See & Spray Ultimate was not significantly influenced by sprayer sensitivity settings. However, cover crop presence did have an impact on their effectiveness, highlighting the importance of integrating cover crops into weed management strategies. These findings suggest that combining multiple approaches, such as targeted spraying systems and cover crops, can enhance overall weed control.

40

Do nitrogen fixation or active carbon ecosystem services differ across cover crop treatments?

H. Frontino[1], E. Rice[1], C. Lowry[1]

[1]*Penn State University, University Park, United States*

Rapid population growth and climate change threats continue to demand greater yields and efficiency from our farms. One sustainable solution is the use of cover crops, which provide several ecosystem services, including erosion reduction, nitrogen (N) retention, carbon (C) sequestration, and N fixation. There is ongoing debate about how cover crop mixtures perform compared to cover crop monocultures. Each approach has its advantages and disadvantages; grass monocultures are more effective for weed suppression, while grass-legume mixtures provide a better balance of weed suppression and N fixation. Mixtures also help reduce the disservices associated with individual species.

In this study, we investigated whether two ecosystem services - active C and biological N fixation - differed between cover crop monoculture and mixture treatments. Nitrogen is a limiting nutrient in terrestrial ecosystems, which motivated our focus on BNF and its potential to increase N availability in cover crop mixtures. A second service of interest is active carbon. Active C refers to the portion of sequestered belowground C that is rapidly broken down and becomes available in the soil. Active soil organic matter is critical for crop nutrient acquisition and is more responsive to management adjustments on a shorter time scale compared to stable nutrient pools.

Our study evaluated several cover crop treatments, including two mixtures (a four-species mixture consisting of triticale (xTriticosecale), crimson clover (*Trifolium incarnatum*), Austrian winter pea (*Pisum sativum*), and canola (*Brassica napus*); and a three-species mixture of triticale, crimson clover, and Austrian winter pea) as well as their respective monocultures (six treatments in total, replicated across four blocks). These species were selected based on their functional diversity and wide use by farmers. Active C was measured across two years using the Permanganate Oxidizable Carbon (POXC) method, while atmospheric BNF was assessed using isotope evaluation of 15N and 14N. Only the BNF data from the spring and fall seasons from one experimental year are presented in this poster. Service outcomes by treatment were evaluated using analysis of variance. Our results showed no significant difference in either service between cover cropping treatments. This suggests that abiotic factors, such as soil texture and temperature, may play a larger role than the cover crop treatments in determining ecosystem service provisioning. However, it is interesting that we did not find a difference between the fallow plot and cover crop treatments. For context, the fallow is normally quite weedy, so those weeds provide an equally good active carbon source as the cover crops. For the service of BNF, while not statistically significant, we did see an increase in N from fixation in the mixtures compared to the monoculture treatments within the legume species.

41

Screening for the response of cereal rye (Secale cereale L.) varieties to allelochemical toxicity in vitro

A. Warris[1], V. Moore[2], C. Reberg-Horton[1], S. Mirsky[3], R. Leon[1]

[1]North Carolina State University, Raleigh, United States, [2]Cornell University, Ithaca, United States, [3]U.S. Department of Agriculture - Agricultural Research Service, Beltsville, United States

The increasing adoption of cereal rye as a cover crop in organic and conventional farming systems can be attributed to its positive impacts on soil conservation, soil fertility, and weed management, which include improved water infiltration, decreased soil erosion, and the production of large amounts of residues that add organic matter to the soil. Cereal rye also suppresses weed germination and establishment through physical barriers on the soil surface, as well as through allelopathic mechanisms. Allelopathy refers to the production and release of chemicals from root exudates that damage seedling tissues, thereby suppressing the emergence and growth of nearby weeds. Although cereal rye varieties have been screened for allelopathic effects using bioindicator species that show injury, the susceptibility of cereal rye plants to injury from adjacent rye plants is not commonly studied. This study examined three commercially available cereal rye varieties—'N.D. Gardner', 'Wheeler', and 'Wrens Abruzzi', which vary in allelopathic potential—to assess the susceptibility of rye seedlings to allelochemical injury from neighboring rye plants and to evaluate the extent of allelopathic interactions among the different rye varieties. Injury was assessed using the *in vitro*, equal-compartment-agar-method and visual estimates of injury to the bioindicator cereal rye seedlings were performed after 10 days. Results from an analysis of variance (ANOVA) revealed no statistically significant differences among the sixteen treatments, indicating that the commercial rye varieties tested were not susceptible to allelochemicals produced by neighboring rye plants. These findings suggest a mechanism in cereal rye that mitigates the effects of allelochemicals from neighboring cereal rye plants, emphasizing the importance of studying this interaction and its implications for weed control in agricultural systems.

42
Investigating Glyphosate-Resistant Goosegrass Populations in New Jersey

C. Mansue[1], A. Butler-Jones[2], W. Bouchelle[1], L. Sosnoskie[2], T. Besançon[1]

[1]Rutgers University, New Brunswick, United States, [2]Cornell University, Geneva, United States

Herbicide-resistant weed biotypes are increasing in numbers. As of March 2021, there are over 196 weed biotypes that are being reported resistant towards certain site of actions in herbicides. Goosegrass (*Eleusine indica*), a problematic summer annual weed, has previously been reported to exhibit resistance to glyphosate in Tennessee and Mississippi. A population of goosegrass from a soybean field in Salem County, NJ, exhibiting poor control with glyphosate was evaluated in a controlled greenhouse study. The putative glyphosate-resistant biotype, designated NJ-Soy, was compared to two putative glyphosate-susceptible biotypes: NJ-Blue, collected from a blueberry field in Atlantic County, and NJ-Turf, collected from a turfgrass site in Middlesex County. A known glyphosate-susceptible biotype was also included, obtained from the USDA National Plant Germplasm collection (USDA). The study aimed to elucidate the physiological mechanisms underlying glyphosate resistance in New Jersey biotypes identified as resistant in dose-response experiments. Results demonstrated that, at 21 days post-treatment, glyphosate applied at the labeled rate of 420 g ae ha^{-1} achieved only 6% control of the NJ-Soy biotype, whereas control of the susceptible biotypes exceeded 85%. Even at a higher application rate of 20,200 g ae ha^{-1}, NJ-Soy exhibited only 67% mortality, markedly lower than the 100% mortality observed in susceptible biotypes. Dose-response analysis revealed that the ED50 (effective dose required for 50% biomass reduction) for the NJ-Soy biotype was 888 g ae ha^{-1}, compared to 59–93 g ae ha^{-1} for the NJ-Blue, NJ-Turf, and USDA biotypes. These findings confirm that the NJ-Soy biotype demonstrates resistance to glyphosate. Ongoing collaborative research with Dr. Caio Brunharo at Penn State seeks to further characterize the resistance mechanisms in this glyphosate-resistant goosegrass population. Insights from this work will contribute to a deeper understanding of resistance evolution and inform the development of more effective weed management strategies.

43

Effect of Cover Crop Termination Timing and Herbicide Strategies for Weed Suppression in Corn

P. Bajwa[1], V. Kumar[1], H. Scatena[1], M. R. Ryan[1], A. Ditommaso[1]

[1]Cornell University, Ithaca, United States

Weeds are the significant pests, causing an average 50% yield loss in corn, valued at $26.7 billion annually in the US. In corn production, 96% of the planted area has received herbicide applications. This over-reliance on specific herbicides has contributed to the development of herbicide-resistant weed populations, complicating effective weed management. Integrated weed management (IWM) strategies, such as cover crops in combination with herbicides, are crucial for improving weed suppression and improving overall weed control efficacy. However, limited knowledge exists on how cover crop termination timings and herbicides affect weed suppression and corn productivity in Northeast U.S. cropping systems. Therefore, this study aimed to investigate the impact of cover crop (mixture of volunteer wheat, red clover and winter peas) termination timings combined with herbicide strategies on weed suppression and corn productivity in Northeast U.S. A field experiment was conducted at a grower field, with corn planted on May 17 during summer of 2024 in Waterloo, NY. Experiment was randomized in a blocked split-plot design involving termination timings [early termination (ET)- May 3 and planting green (PG)- May 20] as main plot factor and herbicides [T1: glyphosate (1261 g ha^{-1}), T2: glyphosate + *s*-metolachlor (1552 g ha^{-1}), T3: glyphosate + *s*-metolachlor + atrazine (1145 g ha^{-1}), T4: glyphosate + *s*-metolachlor + mesotrione (210 g ha^{-1}), T5: glyphosate + *s*-metolachlor *fb* dicamba (280 g ha^{-1}), T6: glyphosate + (*s*-metolachlor + atrazine) *fb* dicamba, T7: glyphosate + *s*-metolachlor + mesotrione *fb* dicamba] as subplot factor. In this study, treatments involving dicamba were applied using a pre-emergence (PRE) followed by post-emergence strategy, while all other treatments were applied as PRE alone. Results showed that early termination of cover crop, mix with glyphosate alone recorded the highest weed density of 43 plants m^{-2} and weed biomass of 171 g m^{-2} across treatment combinations, 13 weeks after planting. This was followed by glyphosate combined with *s*-metolachlor at early termination, which recorded 2-folds lower weed density and 2.5-folds lower weed biomass than control. However, application of soil residual herbicides with glyphosate *fb* dicamba effectively suppressed weeds (>97%), except for glyphosate + *s*-metolachlor *fb* dicamba. Treatments with sequential applications of dicamba reduced the weed density and weed biomass by 8 folds compared to treatments with PRE only, under both termination timings. Cover crops terminated PG accumulated 8% higher corn yield compared to early termination across all herbicide treatments. Among herbicides, glyphosate + *s*-metolachlor + mesotrione *fb* dicamba displayed higher corn yield. In conclusion, cover crop terminated PG significantly improved weed suppression and resulted in better corn yield compared to early termination treatments, and glyphosate + *s*-metolachlor + mesotrione *fb* dicamba effectively suppressed weeds, which led to higher corn yield.

44

Control of Glyphosate-Resistant Waterhemp (Amaranthus tuberculatus) with PRE and POST Herbicides in Enlist Soybeans in New York

H. Scatena[1], V. Kumar[1], L. Sosnoskie[2], P. Bajwa[1], M. Stanyard[3]

[1]Cornell University, School of Integrative Plant Science, Soil and Crop Sciences Section, Ithaca, United States, [2]Cornell University, School of Integrative Plant Science, Horticulture Section, Geneva, United States, [3]Cornell University, Cornell Cooperative Extension, Newark, United States

Glyphosate-resistant (GR) waterhemp has recently been identified in the northeastern United States, including New York. The occurrence of GR waterhemp presents a significant challenge for NY producers, needing the adoption of alternative herbicide strategies for effective management. The main objective of this research was to determine the effectiveness of various preemergence (PRE) and postemergence (POST) herbicide programs for GR waterhemp control in Enlist E3 soybeans. To achieve this objective, a study was conducted during the summer of 2024 in a grower field in Seneca County, NY. The site was in a corn-soybean rotation for >5 yr and had a significant waterhemp seedbank. An Enlist E3 soybean variety was planted on May 24. Herbicide programs, including PRE alone, PRE followed by (fb) early POST, and two sequential POST applications (early POST and mid POST) were tested in the study. The study was established in a randomized complete block design with three replications. Results from the study indicated that PRE-alone applications of chlorimuron + flumioxazin + metribuzin, imazethapyr + flumioxazin + metribuzin, acetochlor + fomesafen, and s-metolachlor + metribuzin at field-use rates provided 83 to 100% control of GR waterhemp 33 days after PRE. However, the effectiveness of acetochlor + fomesafen decreased to 55% by 20 days after early POST. In contrast, all PRE programs fb an early POST treatment of glufosinate + 2,4-D consistently provided excellent control throughout the season (≥96%) and significantly reduced weed density and biomass of waterhemp. Two sequential POST applications of glyphosate were not effective for managing waterhemp (<60% control). In conclusion, effective alternative PRE fb POST herbicide programs tested in this research can help controlling GR common waterhemp in Enlist E3 soybeans.

45

Impact of Cover Crops and Termination Timings on Weed Management in Edamame (Glycine max (L.))

A. S. Brar[1], M. Flessner[2], B. Zhang[2], M. Reiter[1], V. Singh[1]

[1]*Eastern Shore Agricultural Research and Extension Center, Virginia Tech, Painter, United States,* [2]*School of Plant and Environmental Sciences, Virginia tech, Blacksburg, United States*

Edamame's (*Glycine max.* (L)) demand is increasing in the United States due to its several nutritional benefits. However, most of the demand is being met currently by import from the Asian countries. Various production and marketing constraints have been limiting farmers from adopting edamame at a large scale. One of the major production constraints is the limited number of weed control options as compared to grain soybean. Unlike grain soybean, there are very few herbicides registered for use in edamame currently and there are no herbicide-resistant traits available for edamame. Therefore, the development of integrated weed control tools by combining the use of chemicals as well as non-chemical methods such as cover crops is a must. A study was conducted in a split-split plot design at Virginia Tech's Eastern Shore Agricultural Research and Extension Center to compare weed suppression abilities of five cover crop species: cereal rye (*Secale cereale*), black oats (*Avena strigosa*), crimson clover (*Trifolium incarnatum*), hairy vetch (*Vicia villosa*) and rapeseed (*Brassica napus*). Also, two different cover crop termination timings were tested which comprised of early termination at 4 weeks before planting of edamame and late termination at 1 week before planting of edamame. Two edamame cultivars with different maturity groups were compared which included an early maturity cultivar; Midori giant and a late maturity cultivar; VT Sweet. Crimson clover, being the highest biomass yielding cover crop with an average biomass of 7,560 kg ha^{-1} provided excellent weed control but hindered the emergence of edamame which was reflected in the reduced pod yield. Black oats provided complete weed suppression throughout season at both early and late termination timings. In general, for all cover crop species, late termination timing provided a 49% increase in biomass as compared to early termination, which gave a 29% increase in weed suppression. At 21 days after planting, an excellent overall weed control in the range of 83-99% was observed for black oats at both termination timings and for other cover crops at late termination timing. At 35 days after planting, crimson clover at both termination timings provided a weed control rating of 80-95%, which was similar to that of other cover crops at late termination timing. The effect of cover crops and their termination timings was non-significant for Midori giant, but there was a significant reduction in yield of VT Sweet with late termination treatments of crimson clover and hairy vetch. The results of this study will provide base line information to farmers interested in pursuing large scale edamame production in Virginia.

46

Evaluation of soybean weed management in absence of synthetic auxin herbicides

C. Bradshaw[1], D. Contreras[1], J. Alsdorf[1], R. Argueta[1], W. Everman[1]

[1]North Carolina State University, Raleigh, United States

Dicamba is an effective postemergence herbicide for broadleaf weed control that was used extensively on dicamba-tolerant soybean weed management programs. However, with the loss of dicamba label for over-the-top use on soybeans, growers must rely on other herbicide options for weed control in soybeans. A field study was conducted to investigate the effectiveness of different herbicides labeled for use on soybeans. The study included 15 different preemergence herbicide treatments, followed by postemergence treatments with or without a residual application. Excellent weed control was achieved with all preemergence treatments for Palmer amaranth while more variability in control was observed for tall morningglory. Weed control did not differ for either Palmer amaranth or tall morningglory based on the postemergence treatment used. Weed control was excellent for both Palmer amaranth and tall morningglory 28 days after the postemergence treatments. The results of this experiment indicate that excellent weed control is still possible without the use of synthetic auxin herbicides in soybeans, however, weed species present may dictate which herbicides may be more effective.

47
Evaluating the impact of drift reducing agents (DRAs) on spray applications with Unmanned Aerial Systems.

F. Esmaeilbeiki[1], A. S. Brar[1], D. E. Martin[2], V. Singh[1]

[1]Virginia Tech, Painter, United States, [2]USDA, Colledge Station, United States

Drift-reducing agents (DRAs), combined with advanced technologies like Unmanned Aerial Systems (UAS), have significantly improved herbicide application by reducing off-target movement, optimizing droplet size, and minimizing unintended impacts on nearby crops and ecosystems. This study investigates the effect of Drift Reduction Agents (DRAs) on droplet size, focusing on Volume Median Diameter (VMD) to improve agricultural spraying efficiency. Conducted in two phases, the first evaluated three DRAs Liberate, Interlock, and WeatherGard using water as a control, while the second tested these DRAs combined with Liberty herbicide applied via a DJI T50 drone at 4 GPA and a backpack sprayer (BP) at 15 GPA using Liberty as a control. Droplet data, collected using Water-Sensitive Papers (WSPs) in four replications with 17 WSPs per replication, provided a comprehensive assessment of DRA performance across 64 samples per DRA. The results showed that Drift Reducing Agents (DRAs) significantly enhanced droplet size across different application methods, with distinct Volume Median Diameter (VMD) outcomes. In Backpack applications at 15 GPA with Liberty herbicide, all treatments achieved medium-range droplet sizes (236–340 µm), with "Liberate" producing the largest VMD (321 µm), followed by "Interlock" (304 µm), "WeatherGard" (299 µm), and the control (303 µm). While without Liberty, DRAs had a more pronounced effect, with "Liberate" generating the largest VMD (358 µm), followed by "Interlock" (353 µm), "WeatherGard" (344 µm), and the control (331 µm). These findings confirm the efficacy of DRAs in increasing droplet size, supporting effective and drift-reducing spray applications in both traditional and modern systems.

Gap in pagination due to unavailable paper.

Page 48

49

Herbicide Resistance in Palmer amaranth Across Virginia

M. Viric[1], V. Singh[1], M. Flessner[2]

[1]Virginia Tech, Painter, United States, [2]Virginia Tech, Blacksburg, United States

Title: Herbicide Resistance in Palmer amaranth Across Virginia
Authors: Milos Viric, Vijay Singh, Michael Flessner
Palmer amaranth poses significant challenges for management in soybean, corn, and cotton production systems in Virginia, making it one of the most problematic weeds in these crops. In the fall of 2022, we carried out field surveys aimed at understanding the distribution of this herbicide-resistant species and assessing the potential threat to production fields in the area. Following the collection of 50 samples, they were dried, threshed, and tested in greenhouse for sensitivity to trifloxysulfuron (15.76 g ai ha^{-1}), 2,4D (533.68 g ai ha^{-1}), fomesafen (420.3 g ai ha^{-1}), atrazine (2242.8 g ai ha^{-1}), and glyphosate (867.7 g ai ha^{-1}) at 1X field rate. After the herbicide screenings, dose-response assays were performed on the most resistant Palmer populations using six different rates (0.5, 1, 2, 4, 8, 16X), compared to a susceptible population at five rates (0.125, 0.25, 0.5, 1, and 2X). The experiment was conducted in a completely randomized design with three replications and two experimental runs. Survivors were characterized as highly resistant (0-20% injury) or moderately resistant (21-79%) and susceptible (>80% injury). Results showed a high level of resistance to ALS, EPSPS, PSII, and PPO-inhibitor herbicides and also multiple herbicide resistance. The ratio of GR_{50} values indicated that most resistant population had 14-, 14-, and 47- fold resistance to fomesafen, atrazine, and trifloxysulfuron. GR_{50} value for glyphosate resistant population was 1238 g ha^{-1}, and no susceptible standard was available to compare. All 50 samples collected across the state were resistant to glyphosate. There is a risk associated with these herbicide-resistant Palmer amaranth plants that can spread multiple herbicide resistance to surrounding fields through gene flow and giving rise to hybrid plants. The rising prevalence of multiple herbicide-resistant weeds poses a significant barrier to sustainable crop production in Virginia and complicates integrated weed management tactics.

50

How Does Salt Stress Influence Weed Competitiveness in Coastal Organic Rice Production?

G. Singh[1], B. Ward[1], S. White[1], R. Karthikeyan[1], J. Rohila[2], M. Marshall[1], P. Subudhi[3], M. Cutulle[1]

[1]*Clemson University, Clemson, United States*, [2]*USDA ARS, Dale Bumpers National Rice Research Center, Stuttgart, United States*, [3]*Louisiana State University, Baton Rouge, United States*

Climate change is causing sea levels rise and increasing saltwater intrusion. Saltwater intrusion threatens coastal rice (*Oryza sativa* L.), production in South Carolina (SC), where most common and widely recognized rice cultivar is "Carolina Gold," which is salt-sensitive. Additionally, weed infestation has been a major factor contributing to yield losses in organic rice production, particularly in partial saltwater agroecosystems. Addressing this challenge requires a thorough understanding of the effects of salinity on lowland SC rice cultivars and weed interactions to ensure profitability and effective weed control in organic production systems. In first greenhouse study, we compared the performance of four rice cultivars (Carolina Gold, Santee Gold, Doble Carolina, M202) under various salinity treatments with electrical conductivity (EC) levels (0, 0.2, 0.4, 0.8, 1.6 dS m^{-1}) in weed-free and competition conditions. The results showed that 1.6 dS m^{-1} salinity significantly affected rice plants, leading to reductions in biomass and yield under weed competition. M202 and Doble Carolina showed the higher salt tolerance rate compared to the local cultivars, Carolina Gold and Santee Gold. Second greenhouse study was conducted to determine the effects of salinity on rice-weed growth and competition, involving barnyardgrass (*Echinochloa crus-galli* [L.] Beauv.) and hemp sesbania (*Sesbania herbacea* [Mill.] McVaugh). Treatments included five salinity levels (0, 0.2, 0.4, 0.8, 1.6 dS m^{-1}) and five rice-weed competition scenarios (rice, barnyardgrass, hemp, barnyardgrass with rice, hemp with rice). When comparing the salinity effects on weed growth, we observed a reduction in barnyardgrass growth in competition with rice, while hemp sesbania showed no significant impact at 1.6 dS m^{-1} EC. Third, field experiment was conducted to validate the outcomes of the greenhouse studies and to test wide range of salinity levels (0, 0.8, 1.6, 3.2 and 6.4 dS m^{-1}) on six rice genotypes (Carolina Gold, Santee Gold, Doble Carolina, M202, Jupiter, and JN100), both in weed-free and weed-competition. Tidal water was collected from the adjacent saltwater river using a gasoline pump. Then, tidal and freshwater volumes were applied by monitoring EC in different treatment plots. The findings validated our greenhouse observations, demonstrating that M202, Doble Carolina and JN100 showed the highest salt tolerance as evidenced by lower tissue Na^+/K^+ ratios, higher photosynthesis rates, reduced grain sterility, and less yield reduction under saline conditions relative to Carolina Gold and Santee Gold. Weed competition reduced the rice plant height, number of tillers, panicles, dry root, and shoot weight across the entire salinity gradient. Salinity significantly affected weed growth, resulting in 3-5 times decrease in weeding time at 3.2 dS m^{-1} salinity and 8-10 times decrease at 6.4 dS m^{-1} compared to the control (0 dS m^{-1}). Increasing salinity significantly reduced overall weed biomass ($R^2 = 0.841$), although some weed species exhibited differential responses and greater tolerance to salinity stress. This study's findings contribute to understanding of the responses of SC rice cultivars to saltwater intrusion and weed competition in organic production.

51
Calibration of Yield Loss Predictions Based on Aerial Imagery

A. Goldsmith[1], R. Austin[1], C. Cahoon[1], R. Leon[1]

[1]*North Carolina State University, Raleigh, United States*

When making financial and management decisions for current and future growing seasons, it is imperative for farmers to predict their yield, and yield loss relative to previous seasons. Weeds present during the early growing season can have significant impacts on a growers yield relative weed free conditions. The yield impact of weeds has historically been predicted through density measurements, but through UAS derived aerial imagery we hypothesized yield loss could be efficiently and accurately measured through the relative weed leaf cover. During the 2023 growing season, four locations in North Carolina were planted into maize with several herbicide treatments to create a series of weed/maize leaf cover ratios. Aerial imagery was collected at 15-m and 30-m in RGB and four band multispectral to determine optimum image collection specifications. Initial yield loss predictions were moderate across locations but was improved through the development of a correction factor. No differences in prediction accuracy were seen when comparing flight altitude and sensor type. Through further model optimization and automation, it may be possible for growers to obtain in season yield loss predictions, giving them time to make data driven financial and field management decisions.

52

Progeny fitness of native and invasive Parthenium hysterophorus populations: insights from a multi-state common garden experiment

A. Hurtgam[1], S. Kezar[1], A. Ditommaso[1]

[1] *Cornell University, Ithaca, United States*

Parthenium hysterophorus is a significant global weed known for its aggressive growth, ability to germinate in a wide range of temperatures, and impact on agricultural and natural ecosystems. This weed is known to quickly establish in nitrogen-rich environments in its native and non-native ranges. However, the germinability and vigor of native and invasive populations in relationship to soil nitrogen (N) is not well understood. For both germination and vigor experiments, mature F1 *P. hysterophorus* seed were collected from five global populations grown in various N environments in a common garden experiment established in Ithaca, NY in summer 2025. We hypothesize that increased N availability in the common garden experiment would promote the germinability and vigor of both native and non-native populations. The first experiment was a 14-day germination test conducted in a growth chamber environment at Cornell University, with growth chambers set to 20 and 30°C. Findings from this experiment indicate high nitrogen availability benefited the germinability of Mexico and Pakistan progeny at both 20 and 30°C, while mid- and low-nitrogen levels increased germinability for Australia, Israel, and Texas progeny at both 20 and 30°C. The second experiment evaluated the emergence rate and vigor, measured as plant height and biomass accumulation, of F1 seeds produced under competition with native flora as compared to weed-free conditions. Overall, high N availability was advantageous for the biomass accumulation of Pakistan and Australia progeny produced under competition and Texas progeny produced under no competition. Results of this study demonstrate that *P. hysterophorus* seed produced under competition can have a delayed emergence pattern but, nonetheless, produce similar height and biomass as progeny produced under no competition, meaning competition with native flora can promote the growth rate of native and non-native populations.

53

Evaluating Pre-Emergent Herbicides on Pollinator Transplants

H. Lee[1], C. Cahoon[1], Z. Taylor[1], B. Dean[1], J. DeSanctis[1], J. Forehand[1]

[1]*North Carolina State University, Raleigh, United States*

During the summer of 2024, a study was conducted to evaluate wildflower species tolerance to various herbicides applied PRE-transplant and POST-transplant. The wildflower species evaluated were black-eyed Susan *(Rudbeckia hirta)*, purple coneflower *(Echinacea purpurea)*, Scarlet bee-balm *(Monarda didyma*), and Shasta daisy *(Leucanthemum superbum)*. These species were transplanted and treated on May 30, 2024, at the Central Crops Research Station in Clayton, NC. Treatments were replicated 4 times with a randomized strip plot design. The following herbicides were applied PRE-T (rate in g ai ha^{-1}): pyrithiobac (76), mesotrione (140), fluometuron (1212), sulfentrazone (140), indaziflam (58), a premix of pyroxasulfone and flumioxazin (46 + 34), and fluridone (168). The following herbicides were applied POST-T: pendimethalin (1064), pendimethalin (2128), S-metolachlor (1092), and pendimethalin + *S*-metolachlor (2156). An untreated check was included for comparison. When required by label, crop oil concentrate, methylated seed oil, and non-ionic surfactant were added at a one-half percent, one percent, and one-quarter percent volume-per-volume basis, respectively. Plots were 3 m wide by 9 m long with two treated transplant rows. Treatments were applied using a CO_2 backpack sprayer equipped with Teejet flat-fan AIXR 11002 nozzles, calibrated to deliver 140 L ha-[1]. Wildflower injury was visually estimated one, two, four, and eight weeks after treatment, and plot density was recorded 8 weeks after treatment. When pendimethalin and *S*-metolachlor were applied POST-T, minimal injury and reductions in plant density were observed across all species. For black-eye Susan, acceptable (0-15% injury) PRE-T treatments were indaziflam, sulfentrazone, pyrithiobac, fluridone, and mesotrione. Those herbicides caused little to no injury and minimal reductions in plant density. For Scarlet bee-balm, indaziflam, sulfentrazone, and mesotrione caused little to no injury and minimal reductions in plant density compared to the untreated check. For purple coneflower, indaziflam, sulfentrazone, pyrithiobac, fluridone, mesotrione, and fluometuron caused little to no injury and minimal plot density reduction compared to the untreated check. For Shasta daisy, indaziflam, sulfentrazone, fluridone, mesotrione, and fluometuron caused little to no injury and minimal plot density reduction compared to the untreated check. In conclusion, this data suggests that several herbicide options have the potential to be safely integrated over the top of these wildflower species. Herbicides that appear safe will be recommended for further replicated testing. After further testing, 24(c) labels can be perused so that these products may be incorporated into weed management plans by the NC DOT wildflower program.

54

An Individual-Plant-Treatment System for Evaluating Goosegrass Response to Liquid Nitrogen

Keywords:

J. Romero[1], N. Godara[1], S. Askew[1]

[1]*Virginia Tech, Blacksburg, United States*

Cryogenic liquids, defined as fluids with boiling points below -130°F (-90°C), offer innovative approaches to weed control. Liquid nitrogen (LN), at -196°C (-320°F), has demonstrated potential for managing weeds by freezing and destroying plant tissues upon contact. Although effective in crop systems, its broader adoption has been curtailed by high costs and selectivity issues. Recent advancements in artificial intelligence (AI) and precision spraying technologies, however, provide an opportunity for integrating this method into turfgrass management.

This study pursued two primary objectives: to develop a method for delivering precise doses of LN to turfgrass weeds and to evaluate the efficacy of various LN doses for controlling goosegrass (*Eleusine indica*). Field experiments were conducted at two sites using a randomized complete block design (RCBD) with seven LN treatment levels replicated ten times. Goosegrass at the 4-6 tiller stage was treated with LN volumes ranging from 0.64 to 5.12 ml using a custom-designed LN dispenser. Plant responses were quantified via digital image analysis at 3, 7, 14, and 21 days after treatment (DAT).

The results showed that LN doses of 2 to 5 ml per plant were effective in controlling goosegrass, with significant reductions in green cover evident by 3 DAT and sustained control through 21 DAT at higher doses. Lower doses (below 2 ml) only temporarily reduced green cover, with notable regrowth by 7 DAT. An economic analysis indicated that LN application costs could compete with those of prominent commercial herbicides such as Tribute™ Total, Fusilade® II, and Pylex™. The cost of LN application varied from \$64/ha for on-site production to \$378/ha for small-scale purchases, addressing a 1% goosegrass infestation per hectare.

This research highlights LN's potential as a selective, environmentally friendly alternative for weed control in turfgrass settings, particularly when paired with emerging AI-driven precision application technologies. Future research will evaluate turf quality following targeted weed treatment in managed turfgrass.

55

Assessment of Hairy Vetch Monocultures and Mixtures for Biomass Accumulation and Early Season Weed Suppression in Corn.

J. Beville[1], M. Flessner[1]

[1]*Virginia Tech, Blacksburg, United States*

Recently, some Virginia corn (*Zea mays* L.) farmers have opted to start using a hairy vetch (*Vicia villosa* R.) and cereal rye (*Secale cereale* L.) mixture where the cereal rye is selectively terminated in March. Subsequently, the corn cash crop is planted green into the hairy vetch in April. This system has the potential to reduce fertilizer costs due to nitrogen (N) accumulation in hairy vetch, capitalize on state cover crop subsidies, mitigate planting issues associated with cereal rye, and provide early season weed suppression. Farmers have reported observing these benefits as well as increased hairy vetch biomass accumulation. This system needs further research to fully characterize it. Therefore, the goal of this experiment was to determine if a hairy vetch and cereal rye mixture with the cereal rye terminated in March was superior to hairy vetch alone in terms of biomass production and overall weed suppression. A trial was conducted in Blacksburg and Blackstone, Virginia. These trials consisted of four treatments: an early ('Patagonia') and late ('MT') maturing hairy vetch variety both with and without 'Late Abruzzi' cereal rye. Prior to the corn cash crop, hairy vetch biomass was sampled randomly using 0.25 m^2 quadrats within these plots in December, March, and at planting in April. One month after cover crop termination and corn planting, weed density data were taken at each location using two random 0.25 m^2 quadrats within the center row of corn. Biomass data in December and March indicated a location by cover crop interaction. In general, hairy vetch monocultures had more biomass compared to when grown with cereal rye. At planting, biomass data had location and cover crop as main effects without an interaction. Hairy vetch monocultures resulted in greater biomass (3,023 to 3,162 kg ha^{-1}) compared to cereal rye (2,303 to 2,347 kg ha^{-1}). Location and cover crop were significant main effects without an interaction for N content of hairy vetch. Across locations, Patagonia and MT hairy vetch as well as Patagonia hairy vetch with cereal rye were in the top statistical grouping and had 106 to 127 kg N ha^{-1}. MT hairy vetch with cereal rye resulted in 89 kg N ha^{-1} and was not different than Patagonia hairy vetch with cereal rye. While differences were detected between locations in weed density, no differences were detected due to cover crop for Palmer amaranth (*Amaranthus palmeri* L.), common ragweed (*Ambrosia artemisiifolia* L.), morningglory (*Ipomoea sp.*) (Blacksburg only), large crabgrass (*Digitaria sanguinalis* L.) (Blackstone only), other weeds, and total weeds. Treatments did not affect corn yield, but it was greater at Blacksburg (9,467 kg ha^{-1}) than Blackstone (4,538 kg ha^{-1}). Overall, while cereal rye did not positively influence hairy vetch biomass, N content of cover crops was >89 kg ha^{-1} and weed suppression was similar across cover crops. Therefore, farmers can benefit from increased cost share associated with cereal rye while not negatively influencing N accumulation or weed suppression using this system.

56

Laser Weeding can Reduce Weed Cover and Density and Improve Crop Growth and Yields

K. Bill[1], A. Butler-Jones[2], T. Besançon[3], L. Sosnoskie[2]

[1]*American University, Washington, DC, United States*, [2]*Cornell AgriTech, Geneva, United States*, [3]*Rutgers University, New Brunswick, United States*

New York (NY) is a leading producer of several large-acreage vegetable crops, including peas and beets. While herbicides remain the primary weed management strategy for processing vegetable growers, challenges such as limited herbicide options, herbicide resistance, crop injury risks, regulatory changes, and evolving consumer preferences are prompting interest in non-chemical alternatives. One innovative tool is laser weed control, with Carbon Robotics leading the field. Their LaserWeeder™ systems, equipped with high-resolution cameras and deep learning algorithms, precisely target weeds using 150W carbon dioxide (CO2) lasers. Though primarily used in the Western US, this study evaluated their performance under Eastern US conditions.

The experiment was conducted in 2024 at Cornell AgriTech in Geneva, NY, on Lima loam soil using a split-block design with four replicates. Plots measured 30.5 m by 0.76 m. Dominant weeds included common ragweed (*Ambrosia artemisiifolia*), common lambsquarters (*Chenopodium album*), and purslane (*Portulaca oleracea*). Crops included beets (*Beta vulgaris* 'Ruby Queen'), peas (*Pisum sativum* 'Little Marvel'), and spinach (*Spinacia oleracea* 'SV2146VB'). Weed management tactics included S-metolachlor (Dual Magnum®, hereafter H) applied PRE at 720 g ai ha^{-1}, laser weeding (hereafter CR) alone, H followed by (*fb*) CR, and an untreated check (UTC).

Crops were seeded on June 12; the PRE application was made 1 day later. CR operations were conducted on June 21, June 27-28, and July 10. Weed cover and density and crop emergence and stunting were assessed weekly to biweekly. Aboveground crop and weed biomass were collected 33–40 days after planting (DAP). Data were analyzed using R Studio. Crop, treatment, and their interaction were fixed effects; plot was a random effect. Tukey's HSD test was used to separate means.

CR alone (18 plants m^{-2}) and H *fb* CR (2 plants m^{-2}), were, statistically, as effective as H alone (22 plants m^{-2}) for weed control at the time of biomass harvest. Weed biomass averaged over 1 kg m^{-2} in the UTC plots; comparatively, weed dry biomass was 250 g m^{-2} in the UTC compared to 50, 8 and 2 g m^{-2} in the H, CR, and H *fb* CR treatments, respectively. CR used alone significantly improved crop stand establishment and reduced stunting, particularly in beets and spinach. Early season beet and spinach biomass was 40 to 80% greater in the CR treatment compared to the check. Conversely, where H was included in the treatments, beet and pea biomass were reduced 40% to 80%, relative to the control. Injury was exacerbated by significant rainfall and high temperatures that resulted in soil crusting; beet and spinach emergence was delayed in H and H fb CR plots making the crops more sensitive to adverse conditions. Laser weeding still requires further research to understand optimal use under different environmental and edaphic conditions, as well as for weed species with different plant architectures and growth patterns. Future research should also be conducted using commercial units which have more lasers and faster processing speeds.

57

Exploring Weed And Crop Response to Red Clover Living Mulch in Broccoli Production

C. Mansue[1], W. Bouchelle[2], D. Joseph[3], C. Hooks[4], T. Besançon[5]

[1]Rutgers University, New Brunswick, United States, [2]Rutgers University, Chatsworth, United States, [3]University of Maryland, Chestertown, United States, [4]University of Maryland, College Park, United States, [5]Rutgers University, New Brunswicl, United States

The United States ranks as the third-largest global producer of broccoli. Increased broccoli production over recent decades has potentially exacerbated pest pressures, including herbivorous caterpillars, as well as challenges related to weed management. his study aimed to evaluate the effects of intercropping broccoli with a living mulch on pest incidence, weed density, weed biomass, and crop growth and yield. The experiments were conducted during the fall seasons of 2023 and 2024, with data pooled across both years for analysis. Data was generated over the combined years. Results indicate that living mulch reduced insect pest counts in broccoli plots, although insecticides had the most pronounced effect on pest suppression. In low-management plots, caterpillar counts averaged 6 per 10 plants per plot, whereas high-management plots averaged 1 caterpillar per 10 plants. Weed biomass at harvest was significantly lower in plots with living mulch, averaging 16 g m^{-2} compared to 153 g m^{-2} in bare-ground plots. However, the use of living mulch negatively affected broccoli yield and marketability. Bare-ground plots yielded 95% total production and 90% marketable heads per plot, compared to 73% total production and 62% marketable heads in living mulch plots. These yield reductions are attributed to competitive effects, including shading, from the red clover living mulch. In conclusion, living mulch demonstrates potential for reducing caterpillar pests and weed biomass in broccoli production systems. However, the competitive effects of red clover living mulch significantly impact marketable yield and head quality, warranting further investigation into optimization strategies for its use in integrated pest and weed management systems.

58

Understanding the Impact of Saltwater on Weed Dynamics in Organic Rice Production.

G. Singh[1], B. Ward[1], S. White[1], R. Karthikeyan[1], J. Rohila[2], M. Marshall[1], P. Subudhi[3], M. Cutulle[1]

[1]*Clemson University, Clemson, United States*, [2]*USDA ARS, Dale Bumpers National Rice Research Center, Stuttgart, AR, United States*, [3]*Louisiana State University, Baton Rouge, United States*

Rice (*Oryza sativa* L.), a globally important staple crop, faces significant threats from climate change in coastal regions, including rising sea levels and saltwater intrusion. In South Carolina (SC), rice production has been shaped by 'Carolina Gold,' a prized cultivar renowned for its quality, cultural significance, and role in the region's agricultural heritage. However, 'Carolina Gold,' like other current SC rice cultivars, is highly sensitive to salinity stress, making coastal farming increasingly vulnerable to saltwater intrusion. Weed infestation further increases yield losses in organic production systems, especially in partial saltwater agroecosystems. Addressing these challenges requires understanding the combined effects of salinity and weed interactions on rice cultivars to ensure sustainable and profitable production. A field experiment was conducted at Clemson University Coastal Research and Education Center, Charleston, SC to determine the effects of salinity on rice growth and weed dynamics in coastal SC. Six rice genotypes (Carolina Gold, Santee Gold, Doble Carolina, M202, Jupiter, and JN100) were evaluated under weed-free and weed-competition conditions across salinity treatments with electrical conductivity (EC) levels of 0, 0.8, 1.6, 3.2, and 6.4 dS m^{-1}. Three-week-old rice seedlings were transplanted and weed-free plots were hand-weeded at weekly intervals. Brackish water, collected from nearby saltwater river, and applied to treatment plots starting at late tillering stage to crop harvest. Flow meters were installed in each plot to record the volume of water applied. Salinity levels were monitored periodically using EC meters, alongside measurements of pH, total dissolved solids (TDS), temperature and precipitation. The results revealed significant impacts of salinity on both rice and weeds. M202, followed by Doble Carolina, JN100, and Jupiter, showed lower tissue Na^+/K^+ ratios, higher photosynthesis rates, reduced grain sterility, and less yield reduction under saline conditions relative to Carolina Gold and Santee Gold. M202 showed most salt tolerance, with height, tillers, and yield remaining stable up to 3.2 dS m^{-1}, while these parameters began declining in Carolina Gold as early as 0.8 dS m^{-1}. Weed competition significantly reduced rice plant height, tiller number, panicle count, and biomass across all salinity levels. However, salinity suppressed weed growth, leading to 3-5 times decrease in weeding time at 3.2 dS m^{-1} and 8-10 times decrease at 6.4 dS m^{-1} compared to the control (0 dS m^{-1}). While barnyardgrass (*Echinochloa crus-galli* [L.] P. Beauv.) populations decreased significantly at 6.4 dS m^{-1} salinity, *Cyperus* spp. (*Cyperus iria* L. and *Cyperus esculentus* L.) were less affected, showing greater tolerance to higher salinity levels. These findings contribute to the understanding of how salinity and weed competition interact to influence rice production in coastal agroecosystems. Promising salt-tolerant genotypes identified in this study are being incorporated into breeding programs to introduce salt tolerance traits into historically significant cultivars such as Carolina Gold.

59

Effect of Rainfall Timing on Herbicide-Coated Fertilizer Efficacy.

B. Dean[1], C. Cahoon[1], Z. Taylor[1], H. Lee[1], C. Roberts[1]

[1]*North Carolina State University, Raleigh, United States*

In recent years, residual herbicide-coated fertilizer has gained interest across many agricultural production systems. Among the few residual herbicides registered to be applied coated on fertilizer, pyroxasulfone and pendimethalin are two of the most commonly used. Although the efficacy of pyroxasulfone- and pendimethalin-coated fertilizer has been extensively studied, the effect of rainfall timing on these herbicides when applied coated on fertilizer compared to liquid broadcast application remains unstudied. To address this, a greenhouse study was conducted in 2024 at North Carolina State University. The treatment structure was a 2 × 2 × 5 factorial arranged in a randomized complete block design with five replications and two runs. Treatment factors included two herbicides: pyroxasulfone (118 g ai ha^{-1}) and pendimethalin (1,065 g ai ha^{-1}); two application methods: coated fertilizer or liquid broadcast, and five activating rainfall timings: 0 days after application (DAA), and 1, 2, 3, and 4 weeks after application (WAA). At each rainfall timing, 1.9 cm of rainfall was delivered in 0.6 cm increments to prevent flooding. After application, all pots were sub-irrigated to prevent herbicide leaching. Each herbicide was coated on granular ammonium sulfate (AMS) fertilizer and top-dressed at 280 kg ha^{-1} using a plastic container with 4 mm holes in the lid. Liquid broadcast applications were applied using a CO_2-pressurized backpack sprayer calibrated to deliver 140 L ha^{-1}, using AIXR 11002 nozzles. Herbicide-coated fertilizer treatments were compared to a check that received the equivalent rate of non-herbicide treated AMS (280 kg ha^{-1}), while liquid broadcast applications were compared to a completely non-treated check. Applications were made to 20 cm pots filled with a loamy sand soil containing 1% humic matter and pH 6.0. Each pot was planted with 30 Palmer amaranth (*Amaranthus palmeri* S. Watson) seeds then lightly dusted with soil. Preliminary germination tests estimated 70% germinability. Prior to treatment applications, all potting soil was kept dry to prevent premature seed germination. To replicate a field scenario, all pots for each treatment were randomized within a 28 m^2 area, with each treatment evenly distributed across the entire area. For data collection, emerged Palmer amaranth plants were counted 7, 14, 21, and 28 days after treatment (DAT). At 21 DAT, pyroxasulfone was more effective than pendimethalin, resulting in 74% and 56% fewer plants compared to the check, respectively. Each herbicide applied coated on granular AMS (74%) was more efficacious than when applied as a liquid broadcast application (55%). Averaged over herbicides and application methods, plant density reduction was less when treatments received activating rainfall 4 WAA (40%) compared to 0 (73%), 1 (79%), and 2 (69%) WAA. There was no evidence that herbicide coated fertilizer improved density reduction when rainfall was delayed compared to liquid broadcast applications.

Gap in pagination due to unavailable paper.

Page 60

61

Above and Below-Ground Competition Effects on Waterhemp (Amaranthus tuberculatus) Recruitment Dynamics.

L. Wellman[1], J. Wallace[1]

[1]*Pennsylvania State University, State College, United States*

Cover crops are an important tool for integrated weed management but weed suppression potential from cover crops is highly variable. Existing research has focused on identifying cover crop biomass thresholds for sufficient weed suppression, but more research is needed to determine the mechanisms underlying the variation in these data. Belowground resource preemption by cover crops may be an important, non-biomass mechanism of weed suppression that may be impacted by cover crop management. Additionally, density-dependent weed population regulation may further confound the effects of cover crops on weed suppression. A 'safe site' is a microsite that is free from hazards and is capable of supporting one seedling. Quantifying the number of 'safe sites' using seedling recruitment curves accounts for weed density-dependent processes and comparing 'safe sites' across different cover crop management practices can inform weed management. To investigate the effects of cover crop density on resource preemption, and the effect of cover crop management on weed seedling recruitment, we initiated a two-factor randomized incomplete block design in the fall of 2023. Cover crop treatments included a no-cover control, low-density belowground competition (60 kg ha^{-1} rye seed, 19cm rows; harvested), high-density belowground competition (180 kg ha^{-1} rye seed, 14cm rows; harvested), low-density above-and-belowground competition (60 kg ha^{-1} rye seed, 19cm rows; roll-crimped), and high-density above-and-belowground competition (180 kg ha^{-1} rye seed, 14cm rows; roll-crimped). We established five densities of waterhemp (*Amaranthus tuberculatus* (Moq.) Sauer; 25, 250, 2500, 25000, and 100000 seeds $0.25m^{-2}$) within each cover crop treatment. We predicted no change in above-ground resource preemption between rye density treatments, but that high-density rye would allocate more biomass to roots than low-density rye. We also hypothesized that cover crop treatments with the greatest above-and below-ground resource preemption would reduce waterhemp recruitment through a reduction in safe site availability. We assessed rye groundcover in April, and paired above-and belowground biomass in May. Waterhemp demography was assessed prior to rye termination and one month after termination to determine survivorship and recruitment rates. Skellam recruitment functions were fit to waterhemp data to compare the number of safe sites across cover crop treatments. Preliminary results indicate no change in rye above-ground resource preemption, but contrary to expectations, there was an increase in low-density rye root biomass in the upper soil zone within the rye row. The number of safe sites did not differ between cover crop treatments prior to rye termination. Following rye termination, the number of safe sites was lowest in the no-cover control, intermediate in the belowground competition treatments, and not recruitment-limiting in roll-crimped treatments. There was no difference in the number of safe sites with rye density following rye termination. Our preliminary results indicate rye density-mediated biomass allocation did not influence waterhemp recruitment, and in dry summers, the soil moisture conservation of roll-crimped cover crops increased waterhemp safe site availability and protected seedlings from herbivores. This study will be repeated for a second year and the results will be incorporated into waterhemp population prediction models.

62

Weed seed germination and seedling growth in response to allelochemicals in vitro

A. Warris[1], V. Moore[2], C. Reberg-Horton[1], S. Mirsky[3], R. Leon[1]

[1]*North Carolina State University, Raleigh, United States*, [2]*Cornell University, Ithaca, United States*, [3]*U.S. Department of Agriculture - Agricultural Research Service, Beltsville, United States*

Cereal rye (*Secale cereale* L.) is widely utilized as a cover crop due to its potential to improve soil health and control weeds through allelopathic mechanisms. However, the impact of allelochemicals on the germination of weed seeds with varying vigor remains underexplored. The present study examined the allelopathic effects of cereal rye on the germination of *Amaranthus palmeri* L., *Digitaria sanguinalis* L., *Setaria faberi* Herrm., and *Lactuca sativa* L. Seeds were germinated *in vitro* on media containing allelochemicals released by the roots of cereal rye lines with different levels of allelopathic activity, ranging from high to low. Additionally, seeds were aged using the accelerated aging vigor test to alter their vigor. The results indicate that aged seeds experienced a 31% reduction in total germination compared to non-aged seeds. The extent of germination suppression due to allelochemicals varied by species. *Setaria faberi* exhibited a hormetic response when exposed to the low allelopathy treatment, with total germination increasing by more than 20% relative to the untreated control. *Digitaria sanguinalis* showed no response to seed aging, and the high allelopathy treatment resulted in a less than 10% decrease in germination. *Amaranthus palmeri* exhibited the highest total germination and was unaffected by both seed aging and allelopathy treatments. Germination rate was negatively impacted by seed aging, with a lesser effect from allelochemicals. These findings suggest that integrating allelopathic varieties of *S. cereale* into cover crop rotations can be an effective strategy for reducing weed populations. However, the composition of the seed bank, particularly the age structure, may influence the role of allelochemicals in weed emergence.

63

Evaluation of herbicide-coated fertilizer spread over the top of late-season soybeans for Italian ryegrass control

C. Bradshaw[1], B. Dean[1], Z. Taylor[1], C. Cahoon[1]

[1]*North Carolina State University, Raleigh, United States*

Multiple-resistant Italian ryegrass (*Lolium multiflorum* Lam.) has made spring burndown for the planting of summer crops a challenge. Farmers in North Carolina are facing four-way resistant populations of Italian ryegrass that leave no effective method of postemergence control. The lack of effective post-emergent control means we must shift our focus to effective pre-emergent control options. However, Italian ryegrass can emerge before soybeans are harvested, which notes the importance of establishing residual control before soybean harvest. A field study was conducted to determine the effectiveness of residual herbicide-coated ammonium sulfate applied over the top of late-season soybeans to control Italian ryegrass. While ammonium sulfate was not important agronomically, this was used as proof of concept and applications have since been made with muriate of potash. Seven different residual herbicides were applied over the top of soybeans at R6 or R8 growth stage. Clomazone, pyroxasulfone, pyroxasuflone + carfentrazone, and *S*-metolachlor were most effective at controlling fall emerging Italian ryegrass. Clomazone was most effective for controlling spring emerging Italian ryegrass. There was no difference in control between the two application timings. Results indicate that residual herbicide-coated ammonium sulfate applied in late-season soybeans can be an effective method of controlling Italian ryegrass, while some herbicides are more effective than others.

64

Detection of MTI Herbicide Resistance SNP in Goosegrass (Eleusine indica) with Kompetitive Allele Specific PCR (KASP) Genotyping.

K. Tuck[1], M. Elmore[1], E. Patterson[2]

[1]Rutgers University, New Brunswick, United States, [2]Michigan State University, East Lansing, United States

Single nucleotide polymorphisms (SNP)s resulting in amino acid changes can be a source of TSR and targeted with Kompetitive Allele Specific PCR (KASP) genotyping. We developed a KASP assay to detect MTI herbicide tolerance allele resulting in the Leu-136-Phe alpha-tubulin mutation to evaluate the prevalence of the mutation amongst twenty goosegrass (*Eleusine indica*) populations. Ten populations were from golf courses in NJ with historical dithiopyr use, two were from golf courses in PA with historical prodiamine use, and two susceptible turfgrass populations were from NJ where no pre-emergent had been applied. Five row-crop goosegrass populations were also collected from NJ. Sensitivity of goosegrass populations to dithiopyr and prodiamine herbicides were determined with dose-response experiments, conducted in 10 by 10 cm pots filled with 80:20 sand/peat media (v/v). Pots were irrigated and seeded with 50 goosegrass seeds 24 hours prior to herbicide treatment. Rates of prodiamine and dithiopyr (0, 50, 100, 500 and 5000 g/ha) were applied and living seedlings counted 25 DAT. Seedling emergence data were subject to nonlinear regression over herbicide concentration (per label rate) in Prism to generate IC_{50} values for each population.

DNA was extracted from 3-4 untreated plants each population, chosen at random from greenhouse seeded flats seeded with over 100 plants. KASP assays were run in 96-well plates, 4 uL reaction solution (two forward primers (with Hex and Fam tails) and a reverse primer) with KASP master mix added to each well was mixed with 4 uL template DNA for a total reaction volume of 8uL per well. Assays included two negative controls (nuclease free water), two haploid (1:1 mix of each allele) and two positive controls for each allele. Thermocycling conditions consisted of two-step touchdown PCRs with 10 touchdown cycles followed by 30 cycles. Optimal allelic separation was observed at cycle 23 and end point fluorescence data was exported to determine allelic separation for each population.

All five row-crop goosegrass populations, and two turfgrass populations without historical pre-emergent use, were susceptible to dithiopyr (IC_{50} values <0.0001 g/ha). All susceptible populations were wild-type for the Leu-136 allele (no mutation). Three golf course populations had levels of dithiopyr tolerance (IC_{50} values 43, 54, and 13 g/ha) despite lacking the mutation. All eight goosegrass populations with the Leu-136-Phe mutation were from golf course fairways and had various levels of dithiopyr tolerance (IC_{50} values of 29, 181, 39, 45, 47, 37 and 51 g/ha). Two golf course populations were haploids with intermediate levels of dithiopyr tolerance (IC_{50} values 39 and 36 g/ha).

Our assay was the first to detect the Leu-136-Phe amino acid change in non-herbicide treated goosegrass populations. Some wild-type populations lacking this mutation still had intermediate levels of dithiopyr-resistance, but no mutant populations were susceptible to dithiopyr. These results confirm association of Leu-136-Phe with MTI tolerance but indicate other SNPs or mechanisms may also be involved in goosegrass tolerance to dithiopyr. More KASP markers should be developed to further investigate additional mutations associated with MTI herbicide resistance.

65
Effective PRE and POST Herbicides for Controlling Glyphosate-resistant Waterhemp in Conventional Corn

P. Bajwa[1], V. Kumar[1], H. Scatena[1], M. Stanyard[2], M. R. Ryan[1], A. Ditommaso[1]

[1]*Cornell University, Ithaca, United States*, [2]*Cornell University, Newark, United States*

Waterhemp (*Amaranthus tuberculatus*) is among the most troublesome summer annual weeds in corn-soybean rotation in the midwestern United States. Widespread evolution of herbicide resistance among field populations further exacerbates its management. More recently, waterhemp populations have been identified in the Northeastern U.S., including New York (NY). Glyphosate resistance has been confirmed in waterhemp populations from NY. Effective preemergence (PRE) and postemergence (POST) herbicides are urgently needed to manage these glyphosate-resistant (GR) waterhemp populations in NY corn production. In this context, two separate field studies (one for PRE and second for POST) were conducted in a grower's field during the summer of 2024 in Wayne County, NY. The field site was under corn-soybean rotation for >5 yrs and naturally infested with GR waterhemp seedbank. Both experiments were conducted in a randomized complete block design with 3 replications. In experiment 1, PRE herbicides, including premixes of *s*-metolachlor, acetochlor, mesotrione, atrazine, and dimethenamid-P + saflufenacil in various combinations were investigated. In experiment 2, POST herbicides, including various HPPD herbicides (mesotrione, tembotrione, topramezone, tolpyralate) alone or combined with dicamba + diflufenzopyr and atrazine were tested. Results showed that non-treated plots had the highest waterhemp density at 875 plants m^{-2}. In experiment 1, *s*-metolachlor controlled waterhemp for 3 weeks after planting (WAP) before emergence resumed, and *s*-metolachlor + atrazine provided extended control to 5 WAP, which dropped to 32% and 48%, respectively by 8 WAP. However, other tank mixes effectively controlled GR waterhemp for 8 WAP with an average 88% control. Waterhemp density (64 plant m^{-2}) and biomass (90 g m^{-2}) with PRE applied *s*-metolachlor alone did not differ significantly from nontreated weedy check. Premix of *s*-metolachlor + atrazine reduced the waterhemp biomass by 21-folds compared to nontreated weedy check. While *s*-metolachlor + atrazine + mesotrione significantly reduced the waterhemp density by 24-folds and biomass by 90-folds compared to *s*-metolachlor alone. In experiment 2, combining HPPD herbicides with dicamba + diflufenzopyr and atrazine enhanced waterhemp suppression by 7 folds compared to HPPD alone. Mesotrione, tembotrione, tank-mix of tembotrione and dicamba + diflufenzopyr, recorded on average 19-folds lower density than combined control and bromoxynil treatments but 16-folds higher than other treatments tested. Bromoxynil showed significantly similar waterhemp density with control. In conclusion, effective herbicides identified in this research, including s-metolachlor, atrazine, and mesotrione applied PRE and combination of POST applied dicamba + diflufenzopyr, atrazine with HPPD herbicides, can help in controlling GR waterhemp in conventional corn.

66

Determination of the Spatial Relationship between Cover Crops and Summer Annual Weeds

A. Goldsmith[1], A. Dobbs[1], R. Leon[1]

[1]*North Carolina State University, Raleigh, United States*

Grass/Legume cover crop mixtures can provide a series of benefits to agricultural soils such as nutrient scavenging, runoff and erosion prevention, nitrogen fixation, and weed suppression. Cover crop performance is often variable within fields, with their benefits being largely dependent on biomass accumulation. The goal of this study was to determine the spatial relationship of predicted cover crop performance and summer annual weed escapes. It was hypothesized that areas predicted to have high cover crop biomass accumulation would have low levels of weed escapes, while areas of low cover crop biomass would have higher levels of weed escapes. During Spring 2024 a Rye/Legume cover crop prediction model based on structure from motion pixel counts was developed and validated at separate field sites in Maryland and North Carolina. The prediction model was compared to a UAS derived NDVI map of the same field once cover crops were terminated and summer annual weeds had emerged. Little spatial relationship between predicted cover crop biomass and summer annual weed escapes was observed. This could be due to other factors of the field, such as soil variability. For future work we would like to test the spatial relationship over a larger scale to reduce the impact of small weed cohorts.

67
Influence of Cover Crop and Preemergence Herbicide Program in the Effectiveness of See & Spray™ Ultimate

D. Contreras[1], B. Pendleton[1], C. Bradshaw[1], J. Alsdorf[1], R. Argueta[1], C. Cahoon[1], L. Lazaro[2], W. Everman[3]

[1]North Carolina State University, Raleigh, United States, [2]Blue River Technology, Santa Clara, United States, [3]Iowa State Univeristy, Ames, United States

Weeds are among the most detrimental pest groups in agricultural systems, causing significant yield losses and increasing management costs. To effectively combat weed pressure, a proactive approach that integrates multiple weed management strategies is essential. Preemergence herbicides play a critical role in mitigating early-season yield loss by providing effective residual weed control. When combined with multiple modes of action (MOAs), preemergence herbicides offer extended weed suppression. Additionally, cover crops can reduce weed germination by modifying environmental conditions, with cereal rye proving particularly effective in suppressing summer annual weeds like Palmer amaranth (Amaranthus palmer S.) Precision agriculture technologies, such as See & Spray™ (S&S), provides new opportunities for targeted weed control. The S&S Ultimate platform, equipped with dual tanks, enables simultaneous broadcast residual application alongside targeted herbicide sprays, reducing herbicide usage and improving efficiency. A study was conducted to evaluate the effectiveness of weed control and herbicide savings of the S&S Ultimate system, considering the influence of cover crops and the strength of preemergence herbicide programs (single MOA or a premix of multiple MOAs). A 2x3 factorial design was used, consisting of Factor A (cover crop presence or absence) and Factor B (single MOA, premix, or no herbicide), with two untreated checks (cover crop and no cover crop). Herbicide applications were made using an Agronomy Test Machine (ATM), a condensed version of the S&S Ultimate platform, at preemergence, early postemergence, and mid-postemergence timings. Weed control data were collected at each postemergence application and 14 days after mid-postemergence. Results indicated that the presence of a cover crop suppressed Palmer amaranth emergence, with enhanced control when using preemergence herbicides. Preemergence herbicides significantly reduced herbicide output at early postemergence applications when using See & Spray technology, regardless of the herbicide strength or cover crop presence. No differences in Palmer amaranth control were found at mid-postemergence and 14 days after mid-postemergence applications, however there was a significant interaction of cover crop and preemergence treatment used in total volume sprayed. In the absence of a cover crop, spray volume increased were no preemergent herbicide was used. These results highlight the importance of preemergence herbicides for early-season weed control and importance of the use of cover crops, and emphasize how they can be integrated with See & Spray technology.

68

Measuring pollinator exposure to fluorescent pigment on weedy flowers following deterrent treatments

N. Godara[1], J. Romero[1], A. Del Pozo[1], S. Askew[1]

[1] *Virginia Tech, Blacksburg, United States*

Previous research showed that deterrents such as herbicides and UV-altering turf protection products can reduce bee visitation to weedy flowers in turf. Since weedy flowers in turf are the predominant food resource for urban pollinators, such deterrents can be helpful to prevent bee exposure to harmful insecticides. However, it remains unclear if reduced bee visitation, observed over short daily intervals, correlates with reduced exposure to dislodgeable floral residues. To effectively separate various deterrents, a method to assess actual bee exposure to flowers is needed. In 2024, experiments were conducted in Blacksburg, VA, to quantify the efficacy of selected deterrents utilizing fluorescent powder (ECO17 Saturn Yellow®, DayGlo® Pigments) as a stand-in for insecticides to measure honeybee exposure on white clover. Three treatments were compared: mowing on the day of treatment, application of an herbicide mix (2,4-D, MCPP, dicamba; Trimec Classic at 1.52 kg ai ha^{-1}) two days prior, or an untreated control. Using a randomized complete block design with six temporal blocks spaced by approximately one month, and plots separated by at least 6.4 km, we ensured distinct bee exposure conditions. Fluorescent powder was extracted from 1,560 bees using a 9:1 water:dimethylformamide solution and measured via fluorometric analysis. Resulting data on exposed bees and bee powder concentration were subjected to ANOVA and Fisher's Protected LSD at $\alpha = 0.05$. At four hours post-treatment, 80% of bees in the control group were exposed to the powder, dropping to 37% in mowed plots and only 14% in plots treated with herbicides. The powder concentration on bees was significantly lower, with reductions of 80% from mowing and 97% from herbicide use compared to the control. These results underscore the superior deterrent effect of herbicides over mowing in turf settings, significantly reducing bee interaction with treated flowers and thus exposure to potential turf protection chemicals.

69

Short Corn: A Viable Weed Management Tool?

N. Connors[1], J. Wallace[1], H. Karsten[1]

[1]*Pennsylvania State University, University Park, United States*

Dairy farmers in Pennsylvania are converting more acreage to double cropping systems, where they grow corn (*Zea mays)* for silage every year followed by a winter small grain. These corn silage systems present some challenges for weed management: many herbicides are off-label, high nutrient inputs favor nitrophilic weed species, and atrazine has reduced efficacy due to enhanced degradation from repeated application. To address these weed management challenges while maintaining or increasing forage yield, growers are interested in growing corn at higher densities, reducing row spacing, and utilizing new hybrids. Previous research has shown that high density, narrow row planting can reduce light availability to weeds and reduce weed biomass and seed production. However, in corn, these practices often led to lodging which reduced yields. New short corn hybrids are bred to have reduced lodging risk and may allow for higher planting densities. We hypothesized that using new short corn hybrids at high densities with narrow row spacing could increase crop yield and improve weed suppression. Specifically, we hypothesized that in both short stature and standard hybrids, increased corn population and narrow row spacing would each increase corn yield but the effect would be greater when both practices are combined. We also expected both management practices to increase weed suppression. In a field experiment completed in the 2024 growing season, we planted standard and short stature corn hybrids at 79,000 seeds ha^{-1} and 118,000 seeds ha^{-1} in combination with narrow (38 cm) rows compared to standard (76 cm) rows using a factorial treatment structure. In our first year, narrow rows increased corn yield, and at high densities silage yield did not differ between short stature and standard corn hybrids. Additionally, weed suppression was improved for most species with narrow row corn at high densities. Preliminary results suggest that combining new corn hybrids with cultural management practices that enhance weed suppression could allow growers to effectively control weeds with reduced herbicide inputs.

70
Distribution of Herbicide Resistant Palmer amaranth in North Carolina

J. Alsdorf[1], D. Contreras[1], R. Argueta[1], C. Bradshaw[1], C. Cahoon[1], K. Jennings[1], W. Everman[1]

[1]*North Carolina State University, Raleigh, United States*

North Carolina farmers are constantly dealing with competition from various weeds on a yearly basis. The most problematic of these weeds has become Palmer amaranth (*Amaranthus palmeri*), which has been proven to significantly reduce crop yield when gone uncontrolled. Another rising issue among growers is the evolution of herbicide-resistant weed populations. One statewide and two regional surveys for Palmer amaranth populations in 2010, 2005, and 2015, have been previously conducted. These historical surveys have been helpful in understanding the distribution of known herbicide-resistant weed populations. Updated surveys are needed to track the spread of these resistant populations, along with being used to detect putative-resistant populations previously unconfirmed in the state. In the fall of 2022, 137 samples of Palmer amaranth were collected from growers' fields and were screened to determine the efficacy of herbicides: 2,4-D (1064 g ae ha^{-1}), dicamba (562 g ae ha^{-1}), fomesafen (280 g ai ha^{-1}), and glufosinate (596 g ai ha^{-1}). No resistance was found in 2,4-D, dicamba, or glufosinate. Reduced control (<50% control) was found, with 40 populations with fomesafen. In a separate study, five populations were subject to a dose response to confirm resistance to fomesafen. Two susceptible (Bertie and Stanly) and two putative-resistant (Robeson 1 and Robeson 2) populations were selected from the statewide survey, as well as a third putative-resistant population (Nash) collected from a grower complaint. Applications were made to individual plants 5-10 cm tall, at rates of 1/16, 1/8, 1/4, 1/2, 3/4, 1, 2, 4 and 10 times the recomended rate (280 g ai ha^{-1}). Plant injury 21 DAT revealed I_{50} values of 123 and 115 g ai for Bertie and Stanly populations respectively, and 613, 1213, and 2800 for Nash, Robeson 1, and Robeson 2 populations respectively. Ratios of resistance values to susceptible values confirm resistance of Palmer amaranth to fomesafen in North Carolina. Results will be used as extension tools to help growers and agents make informed decisions when choosing a weed control program.

71

Measuring Length of Residual Control as Influenced by Herbicide and Cover Crop Using See & Spray™ technology

R. Argueta[1], D. Contreras[1], B. Pendleton[1], C. Bradshaw[1], J. Alsdorf[1], C. Cahoon[1], L. Lazaro[2], W. Everman[1]

[1]*North Carolina State University, Raleigh, United States*, [2]*Blue River Technology, Santa Clara, United States*

Weed resistance to herbicides necessitates the adoption of new management strategies by growers. Advances in technology can effectively control weed populations. The objective of this study was to determine and compare the duration of residual control provided by various pre-emergence herbicides, both in the presence and absence of a cereal cover crop. The experiment was conducted at the Caswell Research Station in Kinston, North Carolina, where cereal rye (Wrens Abruzzi) was planted and terminated two weeks before planting soybeans. A total of five individual herbicides and five premix herbicides were applied to soybean fields, both with and without the cover crop. Utilizing an agronomy test machine and a John Deere 5100M tractor traveling at 8 mph, a total of 22 treatment combinations were evaluated. Weed density and size were measured at 2, 3, 4, 5, 6, and 7 weeks after treatment. The data was subjected to analysis using ANOVA with SAS® 9.4 software. This study aimed to determine and compare the duration of residual control of several pre-emergence herbicides with and without the presence of a cereal cover crop. The interaction between cover crops and herbicides was found to impact interactions for different weed control efficacy.

72
Glyphosate resistance characterization in Lolium multiflorum from New York

A. Veron[1], M. Stanyard[2], V. Kumar[2], C. Brunharo[1]

[1]The Pennsylvania State University, University Park, United States, [2]Cornell, Ithaca, United States

Lolium multiflorum L. (Italian ryegrass) is a common weed species spread across the US. Glyphosate is the main tool used for Italian ryegrass control; however, cases of resistance have been reported in different states. A suspected glyphosate-resistant population from New York was assessed in this study for its resistance level and mechanisms. We tested the hypothesis that target site and non-target site mechanism have evolved. Four sets of experiments were conducted. First, a dose-response experiment was conducted to quantify the resistance level by treating the suspected-resistant with a known susceptible population with various glyphosate rates (ranging from zero to eight times the field rate of 2.32 a. e. kg ha^{-1}), and applying log-logistic models. Second, the glyphosate target site (*EPSPS*) was sequenced to verify the presence of resistance-endowing mutations at position 106. Third, glyphosate translocation was assessed by quantifying the herbicide concentration at each of treated leaf, culm, or roots 6, 24, 48, and 72 hours after treatment with an HPLC-MS/MS. Lastly, *EPSPS* gene copy number and expression were quantified with gDNA and RNA using real-time quantitative PCR and *ALS* as the housekeeping gene. Dose-response analysis indicated the resistant population required 35 times more glyphosate than the susceptible population to cause 50% growth reduction. Amino acid changes in *EPSPS* and reduced glyphosate translocation were not identified. We observed large increases in *EPSPS* copies in the resistant population ranging from 100 to 800 times compared to the known susceptible, and a positive correlation between *EPSPS* copies and expression. Our results suggested increased *EPSPS* copy number is the resistance mechanism in the Italian ryegrass population studied. Future work will evaluate the mechanism of copy number increase by using long-read sequencing technologies.

73

2024 Survey Results of the Most Common and Troublesome Weeds in Aquatic and Non-Crop Areas

L. Van Wychen[1], J. Miranda[2], S. Chu[3]

[1]Weed Science Society of America, Alexandria, United States, [2]Oregon State University, Corvallis, United States, [3]Texas A&M University, College Station, United States

The 2024 Weed Survey for the U.S. and Canada surveyed the most common and troublesome weeds in: 1) irrigation and flood control; 2) lakes, reservoirs, and rivers; 3) ponds; 4) forestry; 5) natural areas: parks, wildlife refuges; 6) ornamentals: field nursery crops, outdoor containers, Christmas trees; and 7) rights-of-way: railways, roads, public utilities. Common weeds refer to those most frequently seen while troublesome weeds are the most difficult to control, but not necessarily widespread. There were 139 total survey responses from 38 U.S. states. No responses were submitted from Connecticut, Hawaii, Iowa, Maine, Massachusetts, New Hampshire, New York, Pennsylvania, Rhode Island, Tennessee, Vermont, or West Virginia. In addition, no responses were received from any Canadian province. The two most common weeds in irrigation and flood control were 1) Typha spp. followed by Lemna minor, while the two most troublesome weeds included Myriophyllum spp. followed by Hydrilla verticillata. The top three most common weeds in lakes, reservoirs, and rivers were 1) Myriophyllum spicatum, 2) Hydrilla verticillata, and tie for third between Potamogeton crispus and Typha spp. The top three most troublesome weeds in lakes, reservoirs, and rivers were 1) Hydrilla verticillata; 2) Myriophyllum spicatum, and 3) Nitellopsis obtuse. In ponds, the most common and the most troublesome weed was Najas guadalupensis. In forests, the most common weed was Ligustrum spp. and the most troublesome weed was Microstegium vimineum. In natural areas, including parks and wildlife refuges, the most common weed was Cirsium arvense, while there was a 3-way tie for the most troublesome weed that included Bromus tectorum, Cirsium arvense, and Oncosiphon piluliferum. In ornamentals, including field nursery crops, outdoor containers, and Christmas trees, the most common weed was Cardamine spp., while the most troublesome weed was Euphorbia spp. In rights-of-way, including railways, roads, and public utilities, the three most common and troublesome weeds were Sorghum halepense, Bassia scoparia, and Cirsium arvense. The 2024 weed survey data is available at: www.wssa.net/wssa/weed/surveys.

74
Changes in vascular plant diversity, Cuttyhunk, Massachusetts

R. Stalter[1]

[1]*St. John's University, Jamaica,, United States*

Abstract: Cuttyhunk Island, Dukes County, Massachusetts USA, comprising 235 ha, 41 25' N, 70 56' W, was formed during the retreat of the Wisconsin Glacier approximately 14,000 years ago. Our objective was to prepare a complete inventory of the vascular flora of Cuttyhunk Island from 30 September, 2016 to 6 October, 2018, confirming the taxa collected by previous investigators, Fogg (1930) and O'Neill (1974). Of the 364 species collected by Fogg, O'Neill, and Stalter, 107 are invasive taxa and 257 are native species. Invasive taxa have increased comprising 25% of the flora in 1923, 34% in 1974 and 38% in 2018. Across the past 95 years of botanical study, the island's flora has changed in response to a dynamic landscape impacted by human activity, community development, and coastal storms.

75

Eastern US state invasive species policies are a bit of a mess

J. Barney[1], E. Reed[1], S. Salom[1], B. Brown[1], T. Schenk[1], D. Haak[1]

[1] *Virginia Tech, Blacksburg, United States*

Invasive species critically threaten economies, native biodiversity, ecosystem function, and human and animal health. A strong regulatory framework is key to mitigating the threat of current invasive species and preventing new damaging introductions. However, the United States lacks meaningful federal laws and regulations on invasive species, leaving it to individual states. We reviewed all invasive species related laws and regulations for 21 eastern US states to evaluate their taxonomic, pathway, and habitat coverage. We identified 706 policy chapters covering 1200 unique species, of which 523 were identified as introduced/invasive. Policies covering plants and vertebrates outnumbered ones covering invertebrates, though this varied among states. Policies covering terrestrial habitats outnumbered aquatic ones, though it was more balanced in coastal states. Overall there was little overlap or consistency among adjacent states. We encourage states to coordinate efforts at the regional level to better address invasive species issues.

76

Phenotypic differentiation of native and invasive Parthenium hysterophorus populations: insights from a multi-state common garden experiment

S. Kezar[1], M. Cutulle[2], M. Bagavathiannan[3], L. Galvin[4], M. Schill[3], P. Gyawali[3], T. Campbell[2], A. Ditommaso[1]

[1]*Cornell University, Ithaca, United States*, [2]*Clemson University, Clemson, United States*, [3]*Texas A&M University, College Station, United States*, [4]*Oklahoma State University, Stillwater, United States*

Parthenium hysterophorus is a weed of global significance, beyond its native range from the southern region of North America to the northern parts of South America. *P. hysterophorus* is known to invade and thrive in nitrogen (N) rich environments within its native and non-native ranges, however, the influence of soil fertility on its growth and competitive ability in an agronomic environment is yet to be understood. The present study is centered around evaluating the growth and phenotypic differentiation of global *P. hysterophorus* genotypes across different environments, or within the same growing environment, and how various traits may aid demographic success across various agronomic regions. This multi-state common garden study was established in summer 2025 in the historic range (Stillwater, OK, and College Station, TX), the newly invaded range (Clemson, SC), and outside of the known range (Ithaca, NY) of *P. hysterophorus*. To compare various *P. hysterophorus* life histories within each common garden, genotypes were selected from the native (Texas, USA and Texcoco, Mexico) and non-native ranges (Israel, Australia, and Pakistan). The N environment of the common garden consisted of three treatments of spring-applied urea-ammonium nitrate (32% UAN) at high levels (N requirement for 200-bushel corn), medium levels (0.5X N rate), and low levels (no N-applied) based on soil test recommendations. *P. hysterophorus* genotypes were transplanted at the 4 to 6-leaf stage in a Randomized Complete Block Design with four replications of weed-free conditions (no weed competition) and two replications of competition with the native flora for comparison. Growth traits, such as leaf number and plant height, and the phenological development of plants were assessed over a 8-week period. Findings of this ongoing experiment are expected to help develop ecologically-models to predict differences in biotypes, the risk of range expansion or niche shifts, and inform management timings.

77

Weeds in On-Farm Biodiversity Plantings During the First Year of Establishment.

R. Stup[1], S. Westbrook[2], A. Ditommaso[1]

[1]*Cornell University, Ithaca, United States*, [2]*Kansas State University, Manhattan, United States*

Biodiversity plantings are an effective method of improving on-farm biodiversity, However, farmers are often hesitant to place them adjacent to crop fields due to concerns that they may harbor problematic weeds and pest arthropods, or that the intentionally planted species may escape into the crop and act as weeds. The objective of this research was to compare weed and arthropod communities in planted field margins with conventionally managed margins. It was hypothesized that (1) field margin plantings do not increase the abundance of aboveground weeds or pest insects relative to conventional field margin management, and that (2) field margin plantings increase plant biodiversity and beneficial insect abundance relative to conventional field margin management. The margins of corn (*Zea mays*) fields at two sites in New York state were planted with either native wildflowers or grasses, which were then either mowed infrequently or not managed after planting. Conventional field margin management treatments included an infrequent and frequent mowing treatment, glyphosate application at the same rate and time as the corn, and a control with no management operations performed. Measurements included both seedbank and aboveground plant density and diversity, plant biomass, and the abundance of both beneficial and pest arthropods. Planted field margins had greater aboveground plant density than the glyphosate treatment, but density and diversity measurements otherwise did not strongly differ between planted and conventionally managed field margins. We observed no effects of field margin management strategy on both pest and beneficial arthropod abundance. These results indicate that during the first year of establishment, planting beneficial native species in field margins is not likely to increase weed or pest arthropod pressure in the adjacent crop relative to conventional methods of field margin management. Planted margins did not increase biodiversity measures relative to conventional margins during the first year of establishment.

78

Planting Green Effects on Weed Recruitment in a Cereal Rye - Soybean Sequence.

G. Hoffer[1], J. Wallace[1], M. Vangessel[2], N. Basinger[3], A. Hager[4], E. Haramoto[5], J. Lindquist[6], W. Everman[7], E. Law[8], K. Gage[9], E. Miller[9]

[1]*Pennsylvania State University, University Park, United States*, [2]*University of Delaware, Georgetown, United States*, [3]*University of Georgia, Athens, United States*, [4]*University of Illinois, Urbana, United States*, [5]*University of Kentucky, Lexington, United States*, [6]*University of Nebraska, Lincoln, United States*, [7]*North Carolina State University, Raleigh, United States*, [8]*Ohio State University, Columbus, United States*, [9]*Southern Illinois University, Carbondale, United States*

Cover crops are a common weed management tool, but their efficacy is limited by the quantity of biomass produced prior to cash crop planting. Planting green is a cover crop management tactic that could amplify the effects of standard cover crop management by delaying cover crop termination, but these effects are context dependent. This experiment attempted to determine the effects of planting green into a cereal rye (*Secale cereale*) cover crop in soybean (*Glycine max*) cropping systems across the United States. The objectives of this experiment are to (1) model cover crop biomass gains when delaying its termination, (2) evaluate small- and large-seeded species responses, (3) quantify crop and weed management tradeoffs, and (4) quantify cover crop and herbicide interactions. We expected that across locations, planting green would increase cover crop biomass at planting and delay the emergence of small-seeded weed species with little to no soybean yield drag. The experimental design was a 3x2 factorial arranged as a split-plot with four replicates at each site and two sites at each of the nine participating locations spanning multiple soybean-producing states (PA, MD, OH, DE, NC, GA, KY, IL, NE). Main plot treatments were a no cover crop control (No CC), cereal rye terminated 14-21 days before soybean planting (14 DPP), and cereal rye terminated 1-3 days after soybean planting (1 DAP). Split plot treatments were the presence or absence of a PRE application of Fierce XLT (chlorimuron, flumioxazin, and pyroxasulfone mixture). Weed density data was collected 28 days after planting (28 DAP), the R1 soybean growth stage, and at post leaf-drop. Cover crop biomass was collected immediately preceding termination and soybean yields were measure in the middle two rows using a small plot combine. Across all sites, average cereal rye biomass at 14 DPP was 2.9 Mg ha^{-1} and increased 100 kg ha^{-1} for each day of delayed termination. The incidence of Amaranthus weeds 28 days after planting was 47% and 41% less likely in the 1 DAP treatment than the No CC treatment and 14 DPP treatment, respectively. Both the 14 DPP and 1 DAP treatments reduced the likelihood of Amaranthus weeds greater than 10 cm compared to the No CC treatment by 68% and 82%, respectively. Additionally, weeds were 85% more likely to emerge between the 28 DAP and R1 weed censuses in the 1 DAP treatment compared to the No CC treatment. Similarly, large-seeded weeds 28 DAP were 53% less likely to be found in the 1 DAP treatment compared to the No CC treatment. Neither the 1 DAP nor the 14 DPP treatments significantly impacted soybean yield, but the 1 DAP treatment significantly reduced Amaranthus weed densities 28 DAP across all sites. The results of this study suggest that planting green can amplify the benefits of cover crops; biomass generally increases across regions as termination is delayed, weed emergence is delayed or reduced at key management timepoints with fewer off-target weeds, and there is little to no change in soybean yield

Gap in pagination due to unavailable paper.

Page 79

80
Endangered Species Act: A few suggestions on coping with the new label mitigations.

B. Chism[1]

[1]*Chair of WSSA Endangered Species Act Committee, Point of Rocks, United States*

The EPA is assessing the impact of pesticides on endangered species and their critical habitat. This could lead to additional mitigations on pesticide labels to reduce spray drift, runoff, and erosion. This talk will briefly describe the changes coming to labels to help you prepare for these changes.

81

An Integrated Weed Management Program Controls Multiple-Resistant Amaranthus Populations in Corn

L. Smith[1], J. Nunes[2], J. Haarmann[3], M. Kitt[4], D. Belles[4], T. Beckett[4]

[1]Syngenta Crop Protection, King Ferry, United States, [2]Syngenta Crop Protection, Madison, United States, [3]Syngenta Crop Protection, Rochester, United States, [4]Syngenta Crop Protection, Greensboro, United States

Waterhemp (*Amaranthus tuberculatus* [Moq.] Sauer.) and Palmer amaranth (*Amaranthus palmeri* S. Watson) are troublesome weed species in corn production due to the selection of populations with multiple resistance to herbicides commonly adopted for their control. Field trials were conducted over several locations in 2024 both in crop and in bare-ground situations to evaluate herbicide programs for the control of multiple herbicide-resistant *Amaranthus* spp. populations on LibertyLink ® and Roundup Ready ® Corn 2. Results show that integrated weed management programs consisting of preemergence followed by postemergence herbicide applications provided effective multiple herbicide-resistant *Amaranthus* spp. control on LibertyLink and Roundup Ready Corn 2.

82
Evaluating the Redekop Seed Control Unit for Use in Virginia Soybean Production.

E. Russell[1], K. Bamber[1], M. Spoth[1], C. Sias[1], W. Greene[1], M. Flessner[1]

[1]*Virginia Tech, Blacksburg, United States*

Combine modifications for harvest weed seed control, like the Redekop Seed Control Unit (SCU), could have a fit in soybean production. Testing the seed kill of problematic weed species in soybeans is important for determining the efficacy of this mill. Additionally, there are several factors during harvest that could affect a seed impact mills ability to kill weed seeds and reduce its effectiveness. Three ways this is possible are with seeds bypassing the mill by exiting the combine in the straw fraction during harvest, variations in crop yield, and changes in harvest residue moisture, which can impact chaff flow rate and chaff moisture, respectively. This research aimed to determine the seed kill for problematic weeds with a stationary test stand and on the combine, seed loss in the straw during harvest, and how varying chaff flow rates and moisture content would affect the seed kill of the Redekop SCU in soybean. Chaff flow rate and chaff moisture levels tested spanned the range of harvest conditions in soybean. All tested species were killed at >98% with both the stationary test stand and in-field testing. Results also indicated that <3% of weed seeds bypassed the mill in the straw fraction during harvest. Across the range of chaff flow rates tested, a <1% decrease in seed kill occurred across the four species tested. Chaff moisture did not have a significant effect on seed kill. With generally high kill rates (>98%), the ability to kill weed seeds in less-than-ideal harvest conditions (i.e. high moisture chaff), and low seed loss in the straw fraction during harvest, these data indicate that seed impact mills could be used in soybean production to reduce weed seed inputs into the soil seedbank during harvest.

83

Evaluating the Redekop Seed Control Unit for Use in Virginia Wheat Production.

E. Russell[1], K. Bamber[1], M. Spoth[1], C. Sias[1], W. Greene[1], M. Flessner[1]

[1]*Virginia Tech, Blacksburg, United States*

Harvest weed seed control (HWSC) concentrates, removes, or destroys weed seeds as they pass through the combine. Seed impact mills are but one way to implement HWSC. Seed impact mills, like the Redekop Seed Control Unit (SCU), are aftermarket modifications that are mounted directly onto the back of the combine. This mill has the potential to fit within the US wheat production system. To properly evaluate seed impact mill efficacy, testing was conducted to evaluate seed kill of problematic weeds in both controlled and field settings and in less-than-ideal harvest conditions that could affect the mill efficacy. Seed loss in the straw fraction during harvest, which is not processed by the seed impact mill, could affect efficacy. Additionally, any changes in crop yield and harvest residue moisture could impact efficacy as they can alter chaff flow rate and chaff moisture, respectively. This research aimed to determine the seed kill of problematic weed species in controlled and field settings, seed loss in the straw during harvest, and how varying chaff flow rates and chaff moisture affect seed kill of the Redekop SCU in wheat. Chaff flow rate and chaff moisture levels tested spanned the range of harvest conditions in wheat. Results indicated that >89% of all weed seeds of the tested species were killed in both controlled and field settings. Results also indicated that <5% of weed seeds bypassed the mill during harvest in the straw fraction. Chaff flow rate did have an effect on *Lolium perenne* ssp. *multiflorum,* resulting in seed kill decreasing by 3.4% for every 1 kg increase in chaff flow rate. Increasing chaff moisture resulted in seed kill decreasing by 0.43% and 0.015% for every 1% increase in chaff moisture for *L. perenne* ssp. *multiflorum* and *Brassica napus*, respectively, as well. The Redekop SCU delivered high kill rates (>89%) in standard harvest conditions for multiple species, had low seed loss in the straw during harvest, and still killed >82% of seeds in less-than-ideal harvest conditions, which indicates that it has the potential to be implemented in wheat production systems as a tool to reduce weed seed inputs into the soil seedbank during harvest.

84

Cover crop effects on residual herbicide bioavailability: what have we learned?

J. Wallace[1], M. Vangessel[2], M. Flessner[3], C. Smith[1]

[1]Penn State University, University Park, United States, [2]University of Delaware, Georgetown, United States, [3]Virginia Tech University, Blacksburg, United States

Previous research has demonstrated that use of surface mulch can result in reduced soil bioavailability of residual herbicides. Further research is necessary to optimize the benefits of cultural and chemical weed control tactics while minimizing drawbacks. A field experiment was replicated across six site years to investigate cover cropping management effects on the soil bioavailability of residual herbicides. The main plot included alternative cover crop management tactics, including cereal terminated and left standing at the (1) flagleaf stage; (2) heading stage; (3) and anthesis stage, and (4) cereal rye; (5) crimson clover + cereal rye; and (6) hairy vetch + cereal rye that was roll-crimped at cereal rye anthesis. Split plot treatments including application of a test residual herbicide (pyroxasulfone; 0.09 lb ai/ac) applied (1) at the time of cover crop termination (glyphosate 1.12 lb ae/ac; 0 DAT); or (2) as an EPOST treatment (21 DAT). Field assays were employed to measure deposition and washoff processes relative to a no cover crop control. Soil samples were collected 2 to 4 hr after residual herbicide application (deposition assay) and replaced with another set of petri dishes that were collected immediately after an incorporating rainfall (washoff assay). Pyroxasulfone (ppb) was extracted from soil samples and analyzed using HPLC methodology. Additional assays were conducted to measure differences in washoff potential among active ingredients from living (0 DAT) and decomposing (0, 14, 42, 84 DAT) cereal rye residues using simulated rainfall (1.9 cm). Herbicide active ingredients differed in polarity, ionic charge, and water solubility, including: mesotrione, pyroxasulfone, atrazine, s-metolachlor, and pendimethalin. When pyroxasulfone was applied at cereal rye termination in a planting green scenario, soil bioavailability decreased from 86% to 33% as biomass increased from 1 to 8 Mg ha-1, respectively, when cereal rye was left standing. Across this biomass range, approximately 10% of the herbicide that is present in the soil came from the washoff process. In comparison to standing cover crops, soil bioavailability decreased from 81% to 13% in a planting green scenario when roll-crimping prior to application. Grass-legume cover crop mixtures reduced residual herbicide concentration in the soil compared to a grass monocultures. Averaged across biomass levels, pyroxasulfone applied into roll-crimped cereal rye – hairy vetch Compared to planting green into standing cereal rye, delaying application of pyroxasulfone to EPOST increased its concentration by 10% at 4 Mg ha^{-1} cereal rye residue to 100% at 8 Mg ha^{-1}. Washoff potential of alternative soil-residual herbicides with differing properties showed partial support that log K_{ow}, a measure of lipophilicity, can be used to predict washoff potential when applied into a living cover crop. The decomposition stage of cereal rye terminated at anthesis did not affect the amount of herbicide washoff up to 42 days after termination, which is within the residual herbicide application window in corn and soybean management scenarios.

85

Effect of Multiple Very-Long-Chain Fatty Acid-Inhibiting Herbicide Applications on Soybean Yield

W. Everman[1], D. Contreras[2], J. Alms[3], D. Vos[3], E. Jones[3]

[1]Iowa State Univeristy, Ames, United States, [2]North Carolina State University, Raleigh, United States, [3]South Dakota State University, Brookings, United States

Very-long-chain fatty acid-inhibiting herbicides (Group 15) are applied extensively to manage weeds preemergence. Group 15 herbicides are being applied post residual to reduce the amount of later germinating weeds. Minimal data is present to determine the influence of multiple applications of these herbicides on soybean yield. Field studies were conducted to determine the effect of multiple applications of Group 15 herbicides on soybean yield. Experiments were conducted at two locations in North Carolina (Kinston and Rocky Mount) and South Dakota (South Shore and Volga), respectively. *S*-metolachlor (1703 g ai ha^{-1}) and encapsulated acetochlor (1261 g ai ha^{-1}) were applied at three different timings: preemergence, early postemergence (V2-V3), and mid postemergence (V5-V6). All timings were applied in a factorial arrangement. These two herbicides were selected based on immediate plant availability (*S*-metolachlor) and delay availability (encapsulated acetochlor). Glufosinate (655 g ai ha^{-1}) and glyphosate (1260 g ae ha^{-1}) were applied between the early and mid postemergence application to manage weeds that could confound putative yield loss. Experiment location was considered random to increase the inference of the data. The main factor herbicide did not influence soybean yield. The main factor of timing influenced soybean yield where a loss occurred with the early and mid postemergence application. The interaction between herbicide and timing was not significant. These results suggest that multiple applications of Group 15 herbicides do not affect soybean yield.

86

Are There Any Herbicides to Control Weedy Grasses in Spring Oats?

D. Lingenfelter[1]

[1]*Penn State University, University Park, United States*

Farmers struggle to control weedy annual grasses in spring oats (*Avena sativa*) production fields. In Pennsylvania and the Mid-Atlantic region there are no herbicides that are labeled for use in oats that will provide consistent control of a variety of summer annual grassy weeds such as giant and yellow foxtail (*Setaria faberi* and *pumila*), large crabgrass (*Digitaria sanguinalis*), and fall panicum (*Panicum dichotomiflorum*). Some wonder if products like Osprey (mesosulfuron) and PowerFlex HL (pyroxsulam) which provide weedy grass control and are labelled for use in wheat might be an option; while others surmise that products like Prowl H2O (pendimethalin), Dual Magnum (s-metolachlor), Anthem Flex (pyroxasulfone + carfentrazone), or others might be an answer. Thus, a field study was conducted in Centre County, Pennsylvania in 2024 to investigate spring oats safety and giant foxtail control with various herbicides at different application timings. The study was designed as a basic screen to obtain initial baseline data for each of the products and thus was arranged as a strip trial in plots that measured 3 x 27 m. Oats were planted on April 9. Herbicides were applied with a small-plot, CO2-backpack sprayer system that delivered 140 L ha-1 thru TeeJet AIXR110015 nozzles. Herbicides included: pendimethalin (1065 g ai ha-1), pyroxasulfone (105 g), s-metolachlor (1065 g), diuron (1345 g), bicyclopyrone (51 g), pyroxsulam (18 g), mesosulfuron (15 g), and metribuzin (157 g). Depending on the herbicide, treatments were applied as DPRE (5-7 DAP), EPOST (1-2 leaf oats), and/or POST (20-25 cm oats; weeds 2-5 cm). Visual crop injury and giant foxtail control ratings were collected routinely during the season.
Preliminary data showed that pendimethalin applied either DPRE or EPOST caused 0-15% oats injury and 83-90% giant foxtail control at 8 and 12 WAP; pyroxasulfone caused 15-30% oats injury when applied either DPRE or EPOST and 70-85% foxtail control; s-metolachlor caused 5-7% oats injury and 85-90% foxtail control when applied either DPRE or EPOST; diuron applied at all timings caused 7-65% oats injury and 85-95% foxtail control; bicyclopyrone applied at either DPRE or POST caused 0-5% oats injury and 45-83% foxtail control; the remaining herbicides were all applied POST and pyroxsulam caused 50-65% oats injury and 60-70% foxtail control; mesosulfuron caused 65-98% oats injury and 50-70% foxtail control; and metribuzin caused 35% oats injury and 25-40% foxtail control.
In summary, preliminary results indicate that there are herbicides that only cause minimal injury to spring oats yet provide adequate control of giant foxtail. Potential herbicide candidates could include pendimethalin, s-metolachlor, diuron, and bicyclopyrone. Pyroxasulfone might also be a candidate, but the rates might need to be reduced to improve crop safety. Pyroxsulam, mesosulfuron, and metribuzin in general cause unacceptable crop injury and thus should not be considered as viable candidates. Further research on herbicide rates and timings will need to be conducted to determine if the potential candidates could be used successfully in spring oats.

87

Response of Container-grown Hydrangea to Pre-emergence Herbicides.

A. Witcher[1]

[1]*Tennessee State University, McMinnville, United States*

Pre-emergence herbicides prevent weed seed germination and development, and several granular products are registered for use in container-grown nursery crops. Nevertheless, some crop species are sensitive to certain pre-emergence herbicide active ingredients which limit their use. Hydrangea is a popular landscape plant, accounting for 16% of sales (in value) of all deciduous shrubs and second only to rose. Unfortunately, hydrangea is sensitive to many pre-emergence herbicides and only a few pre-emergence herbicides are labeled for use in hydrangea. The limited availability of pre-emergence herbicides labeled for use in hydrangea limits weed management options and growers must resort to manual hand weeding which is labor intensive and time consuming. The objective of this study was to identify pre-emergence herbicides for safe use in three common hydrangea species [*Hydrangea arborescens* L., *H. macrophylla* (Thunb.) Ser., and *H. quercifolia* W. Bartram]. In May, hydrangea liners were transplanted to 5.7 L nursery containers filled with a pine bark substrate. *Hydrangea arborescens* ('Incrediball'), *H. macrophylla* ('Let's Dance Lovable' and 'Penny Mac'), and *H. quercifolia* ('Gatsby Gal' and 'Ruby Slippers') were evaluated. Pre-emergence herbicides were applied 1 day and 8 weeks after transplant. At the first herbicide application, branches were lightly swept to remove loose granules. Herbicides included dimethenamid-P + pendimethalin, dithiopyr + isoxaben, flumioxazin, flumioxazin + prodiamine, oxadiazon + oxyfluorfen, oxyfluorfen + pendimethalin, oxyfluorfen + prodiamine, prodiamine + isoxaben, trifluralin + isoxaben. No phytotoxicity was observed on any of the hydrangea cultivars over the four month period. Four months after the first application, none of the herbicides resulted in a reduction in plant shoot height or width for any of the hydrangea cultivars. Shoot dry weight was reduced by 30% for *H. quercifolia* 'Gatsby Gal' treated with flumioxazin + prodiamine, oxadiazon + oxyfluorfen, prodiamine + isoxaben, trifluralin + isoxaben. Shoot dry weight was not negatively affected by herbicide for any other hydrangea cultivars. Leaf greenness was not affected by herbicide in any hydrangea cultivar. Herbicide granules can fall into crevices in leaves and buds resulting in plant damage. Many of the products in this trial were not labeled for use in hydrangea, yet little to no damage was observed when branches were lightly swept to remove loose herbicide granules after the first application. Additional research is warranted to test these products on additional cultivars of these hydrangea species and to determine effects without manipulating the branches after application.

88

Turfgrass Species and White Clover Inclusion for Crabgrass Suppression.

D. Tuck[1], M. Elmore[1]

[1]*Rutgers University, New Brunswick, United States*

A field experiment was conducted at the Rutgers Horticulture Farm No. 2 in North Brunswick, NJ, to evaluate the effect of turfgrass species and white clover (*Trifolium repens*) inclusion on smooth crabgrass (*Digitaria ischaemum*) encroachment in a simulated lawn. Treatments consisted of seven turfgrass species in a complete factorial with no clover inclusion, 'Dutch White' white clover at 25 kg ha^{-1}, 'Pipolina' Microclover at 25 kg ha^{-1}, or pre-emergence applied prodiamine (Barricade 4FL; 840 g ha^{-1}). Turfgrass species treatments consisted of 'Stellar 4GL' perennial ryegrass (*Lolium perenne*) at 390 kg ha^{-1}, 'Titanium GLS' turf-type tall fescue (*Festuca arundinacea*) at 390 kg ha^{-1}, 'Tenacious' hard fescue (*Festuca brevipila*) at 300 kg ha^{-1}, 'Radar' Chewings fescue (*Festuca commutate*) at 300 kg ha^{-1}, 'Cardinal II' strong creeping red fescue (*Festuca rubra*) at 300 kg ha^{-1}, '007XL' creeping bentgrass (*Agrostis stolonifera)* at 75 kg ha^{-1}, 'Bluenote' compact-type Kentucky bluegrass (*Poa pratensis*) at 122 kg ha^{-1}, and 'Argyle' midwest-type Kentucky bluegrass at 122 kg ha^{-1}. Monostands of 'Dutch White' and 'Microclover' at 50 kg ha^{-1}, respectively, were also included. Plots were established in September 2023 following a glyphosate application on a site with a history of severe crabgrass infestation. Turf and clover seed were applied by hand with a shaker jar and seeding frame to 1.5 by 1.5 m plots arranged in a randomized complete block design.

Following establishment, the site was maintained at a 7 cm mowing height and was irrigated as needed to prevent wilt. Fertilizer was applied at 50 kg N ha^{-1} on a plot-by-plot threshold basis when the turfgrass quality was < 6 due to poor turfgrass density, N deficiency symptoms, or presence of dollar spot (*Clarireedia* ssp.). Smooth crabgrass cover was evaluated visually on a 0 (i.e., no cover) to 100 (i.e., complete coverage) percent scale in 2024. Data were analyzed using the GLIMMIX procedure in SAS ($P = 0.05$) and Fisher's protected LSD test was used to separate means. A turfgrass species-by-clover treatment interaction was detected for crabgrass cover in both July and August 2024.

Plots receiving prodiamine resulted in the lowest crabgrass cover values at both rating dates. In the absence of prodiamine, clover inclusion increased turfgrass quality and decreased crabgrass encroachment compared to turf-only plots. Microclover-included plots typically outperformed White Dutch-included plots, especially in August when crabgrass was most severe. The benefits of clover inclusion were greatest in perennial ryegrass and both cultivars of Kentucky bluegrass. In July, microclover-included perennial ryegrass and both cultivars of Kentucky bluegrass had less than 10% crabgrass cover while PRG and KBG monostands had crabgrass cover > 70%. In the absence of prodiamine and clover, creeping bentgrass, chewings fescue, and hard fescue were the turf species that suppressed crabgrass the most. In July, all plots of these three species were in the lowest statistical category of crabgrass cover (<13%). 'Dutch White' and 'Microclover' crabgrass suppression was poor in the absence of turf. In August, clover monostands had crabgrass cover of over 95%.

89

Ethoxylated Lecithin Formulations Improve Quinclorac Rainfastness on Smooth Crabgrass

S. Askew[1], B. Corbett[2], C. Goncalves[3]

[1]Virginia Tech, Blacksburg, United States, [2]Albaugh Speciality, Raleigh, United States, [3]UC Cooperative Extension, Ukiah, United States

Quinclorac is widely recognized as the leading herbicide for managing smooth crabgrass in turfgrass, yet its efficacy can vary due to factors such as the growth stage of the weed, environmental conditions, and the type of surfactant used. With a low water solubility and a tendency to crystallize, quinclorac's performance benefits significantly from the addition of surfactants. Linear alcohol ethoxylates (LAE) and methylated seed oil (MSO) have been shown to enhance quinclorac's effectiveness against annual grass weeds. Recently introduced, Quintessential herbicide features H-value technology, which includes proprietary lecithin-based surfactants, promising enhanced weed control. Ethoxylated lecithin (EL), an environmentally friendlier alternative to LAE due to its soybean derivation, shares similar amphiphilic enhancements through ethoxylation. This study aimed to assess Quintessential against traditional quinclorac formulations lacking EL across various simulated rainfall scenarios and to compare different EL-containing formulations with Quintessential for smooth crabgrass control. Two experiments were conducted using a rainfall simulator designed to mimic natural rain with 2-3 mm drops at near-rainfall velocity. The first study was repeated 3 times and employed a randomized complete block design with five replicates, evaluating three herbicide treatments (quinclorac, quinclorac + MSO, and Quintessential + MSO) across six rainfall timings (0.5 to 12 hr). The second study was conducted twice and tested six herbicide formulations at 4 and 12 hr rainfall timings each replicated four times. Results indicated a stepwise increase in smooth crabgrass control from quinclorac alone (63%), to quinclorac with MSO (75%), and highest with Quintessential + MSO (91%) across all rainfall scenarios and three site years. The main effect of rainfall time suggested an upper asymptote of smooth crabgrass control beginning at 4 hours rainfastness based on exponential regression curves. In the second study, formulations with EL matched Quintessential's performance at both half and full EL doses, but not when EL was omitted, highlighting EL's contribution to quinclorac's efficacy in managing smooth crabgrass under varied rainfall conditions.

90

Pollinator Plant Response to Preemergence Herbicides.

J. Derr[1], A. Rana[1]

[1] *Virginia Tech / HRAREC, Virginia Beach, United States*

Two trials were established - preemergence herbicides applied at seeding and herbicides applied after annual flower emergence. Pendimethalin (Pendulum AquaCap) was applied at 1.55, 3.1, and 6.2 pt $\cdot A^{-1}$ (0.7, 1.5, and 2.9 lb ai$\cdot A^{-1}$). Metolachlor (Pennant Magnum) was applied at 1.3, 2.6, and 5.2 pt$\cdot A^{-1}$ (1.2, 2.5, and 5.0 lb ai$\cdot A^{-1}$). Imazapic (Plateau) was applied at 2, 4, and 8 fl oz$\cdot A^{-1}$ (0.03, 0.06, and 0.13 lb ai$\cdot A^{-1}$). For treatments applied at seeding, at 16 DAT, compared to the nontreated plots, no rate of pendimethalin reduced stand of marigold, sunflower, cosmos, or zinnia. Metolachlor at all rates reduced stand of marigold, and the highest rate reduced stand of cosmos and zinnia. The two higher rates of imazapic reduced stand of marigold, sunflower, and zinnia and all rates reduced stand of sunflower. The highest rate of pendimethalin reduced zinnia flowering at 41 DAT compared to the two lower rates of pendimethalin. All rates of pendimethalin reduced flower counts of marigold at 54 DAT compared to nontreated plots but no rate reduced flowering in sunflower, cosmos, and zinnia. Metolachlor at all rates essentially prevented flowering in marigold 54 DAT and the highest rate significantly reduced flowering in cosmos and zinnia. Imazapic significantly reduced flowering in marigold and sunflower at all rates and the highest rate significantly reduced flowering in cosmos and zinnia 54 DAT. The best flowering in marigold was with the lowest rate of pendimethalin at 61 DAT. Pendimethalin did not reduce flowering in sunflower, cosmos or zinnia at 61 DAT. Of the herbicides evaluated, pendimethalin posed the greatest safety overall to zinnia, marigold, sunflower, and cosmos.

For treatments applied after annual flower emergence, pendimethalin did not injure marigold, sunflower, cosmos, or zinnia at 13 DAT. Metolachlor injured zinnia at the 2 higher rates. Imazapic significantly injured marigold, sunflower, cosmos, and zinnia. At 30 DAT, pendimethalin did not injure marigold, sunflower, cosmos, or zinnia. Metolachlor did not appear to injure any of the 4 species at this time. Imazapic caused severe injury to sunflower and zinnia, with lower injury to marigold and cosmos. At 30 DAT, pendimethalin reduced flowering in cosmos but did not reduce flowering at the 2 lower rates in marigold, sunflower, and zinnia. Metolachlor did not appear to reduce flowering in marigold, sunflower, cosmos, and zinnia. Imazapic at all rates decreased flowering in all 4 species. Overall, marigold, sunflower, cosmos, and zinnia appear to tolerate pendimethalin and metolachlor applied after crop emergence, especially at the lowest rate tested. Imazapic injured marigold, sunflower, cosmos, and zinnia and caused significant reduction in flowering at the two higher rates in these species.

91
Protecting Pollinators from Isecticide Exposure via Transient Floral Suppression

S. Hale[1], N. Godara[1], J. Romero[1], S. Askew[1]

[1] *Virginia Tech, Blacksburg, United States*

Pollinator decline in urban environments is closely tied to insecticide exposure, with turfgrass weedy flowers serving as crucial food sources. While herbicides can deter honeybees from foraging on weed-infested turf, they also result in prolonged removal of these floral resources. An alternative approach involves using plant growth regulators (PGRs) and low-dose herbicides to temporarily suppress blooms, thereby reducing pollinator exposure to insecticides while allowing for floral recovery. This study, conducted in summer 2024 in Blacksburg, VA, investigated how different PGR and herbicide treatments impact white clover floral density, honeybee behavior, and turfgrass health. We observed a strong positive correlation between white clover bloom density and honeybee visits. Treatments with ethephon and the ethephon-dikegulac combination led to a significant reduction in white clover blooms four weeks post-application, with blooms fully recovering by eight weeks. Conversely, florasulam and high-dose dikegulac treatments maintained bloom suppression beyond eight weeks. These results highlight the potential of transient bloom suppression strategies in MidAtlantic turf management to safeguard pollinators from insecticide exposure in treated areas, facilitating floral resource recovery.

92
Postemergence Broadleaf Weed Control with Individual Auxin Herbicides and Mixtures

M. Elmore[1], A. Patton[2], A. Kowalewski[3], S. Askew[4], D. Tuck[1], B. McNally[2], B. Corbett[5]

[1]Rutgers University, New Brunswick, United States, [2]Purdue University, West Lafayette, United States, [3]Oregon State University, Corvallis, United States, [4]Virginia Tech, Blacksburg, United States, [5]Albaugh LLC, Ankeny, United States

Field experiments were conducted in 2023 (year 1) and 2024 (year 2) to evaluate synthetic auxin herbicides and mixtures for postemergence white clover (*Trifolium repens*) control. Replicate experiments were conducted in Freehold, NJ, West Lafayette, IN, Blacksburg, VA, and Corvallis, OR totaling eight site-years. Treatments consisted of eight herbicide treatments applied singly and a non-treated control. Herbicide treatments consisted of dicamba (124 g ae ha^{-1}), 2,4-D (1330 g ae ha^{-1}), MCPA (354 g ae ha^{-1}), MCPP-p (353 g ae ha^{-1}), 2,4-D + MCPA + dicamba (1340 + 348 + 123 g ae ha^{-1}) and two different formulations of 2,4-D + MCPP-p + dicamba (1340 + 348 + 123 g ae ha^{-1}). All herbicides were dimethylamine salt formulations. Herbicide treatments were applied between 18 May and 22 August in 2023 and 2024 using CO_2-powered booms with 375 to 840 L ha^{-1} of water carrier. All eight sites had natural infestations of white clover mixed with turfgrass species. The NJ, IN, and OR sites were cool-season turfgrasses maintained as a simulated lawn with respect to mowing, irrigation and N fertility. The VA sites were bermudagrass (*Cynodon* spp.) maintained as a low-input golf course fairway. Plots were arranged as an RCBD with four replications. In NJ and VA, white clover control was evaluated visually on a 0 (i.e., no control) to 100 (i.e., complete control) percent scale relative to a non-treated control. In IN and OR, white clover control was assessed and then converted to control based on reductions in pre-treatment cover. Data were analyzed using the GLIMMIX procedure in SAS (P = 0.05) and Fisher's protected LSD test was used to separate means. Location-by-year-by-treatment interactions were significant for both white clover control; data for each location were analyzed separately. Year-by-treatment interaction were significant for NJ and VA locations. Therefore, white clover data for OR and IN were pooled across years, while NJ and VA data were analyzed separately across years. Control assessed six weeks after treatment are presented for all locations.
MCPP controlled white clover more than MCPA at all locations and years and more than 2,4-D at all locations except VA in year 1. Dicamba provided more white clover control than MCPP in five of eight site-years. 2,4-D + MCPA + dicamba provided more white clover control (82 to 96%) than dicamba (61 to 83%) in three of eight site-years, while these treatments provided similar control (83 to 93%) in the other five site-years. All three-way mixtures provided similar white clover control to each other (91 to 100%) in OR, IN, and VA. 2,4-D + MCPP + dicamba provided more clover control (94 to 97%) than 2,4-D + MCPA + dicamba (82 to 86%) in NJ.

93
IR-4 Weed Science Update - Food Crops

R. Batts[1]

[1]*IR-4 Project, Raleigh, United States*

<u>Residue projects</u>
As of Dec 1, 2024, IR-4 data submitted to EPA led to over 920 new specialty crop uses in 2024. Nearly 60 of those were for weed science uses, including new or expanded uses for triclopyr and saflufenacil. Triclopyr was approved for use in sugarcane for control of Divine nightshade (*Solanum nigrescens*). Approvals for saflufenacil include uses for citrus fruits, pome fruits, stone fruits, tree nuts, and mint. IR-4 submitted three herbicide data petitions to EPA in 2024 (saflufenacil, clopyralid and flumioxazin). These submissions could potentially lead to approximately 90 new uses. Twenty-one new herbicide and plant growth regulator (PGR) magnitude-of-residue studies began in 2024. They include clethodim on fig, ethephon on ginseng, ethofumesate on swiss chard, flumioxazin + pyroxasulfone on cucumber, squash and cantaloupe, linuron on mint, stevia and green onion, pyridate on sweet corn, mesotrione on sesame, metribuzin on potato, 1-naphthaleneacetic acid on plum and hazelnut, norflurazon on clover grown for seed, quinclorac on grape, s-metolachlor on carinata and field pennycress, and tolpyralate on hazelnut, blueberry and sweetpotato. Eighteen new herbicide and PGR residue studies are scheduled to begin in 2025. These include bromoxynil on canola, dry pea and sunflower, ethofumesate on pepper, flazasulfuron on hops, glufosinate on caneberry and sugarcane, linuron on basil, dry bean, succulent pea and sweetpotato, orthosulfamuron on pomegranate, phenmidepham on spinach, pyridate on tomato, quinclorac on cherry, peach and plum, and terbacil on peach.
<u>Product Performance projects</u>
Generating Product Performance (efficacy and crop safety) data to support registration of pest management tools in specialty crops continues to be an important and expanding part of the IR-4 annual research plan. The number of on-going herbicide Product Performance studies in 2024 was 28, including over 70 individual trials. The 2025 Performance research plan for herbicides and plant growth regulators includes thirty continuing or new studies (nearly 70 individual trials).
<u>Integrated Solutions projects</u>
IR-4's Integrated Solutions (IS) Program is structured to assist specialty crop growers outside of the traditional single product/single crop residue and product performance research. IS research efforts focus on crop-pest combinations to address solutions in these four areas, 1) pest problems without solutions, 2) resistance management, 3) products for organic production and 4) pesticide residue mitigation. In 2024, there were seven active IS projects with herbicides and plant growth regulators (25 individual trials), two of which will continue in 2025. Two new weed control IS studies are tentatively scheduled to begin in 2025 (~6 individual trials), including avocado and tame mustard.

94

Do Herbicide Safeners Have a Place in Vegetable Weed Managment?

M. Cutulle[1]

[1]*Clemson University, Clemson, United States*

Due to the research and development costs associated with herbicide discovery there are minimal herbicides registered in low-acreage high-value vegetable crops. A potential opportunity to increase the safety of herbicides in specialty crops would be to evaluate natural product herbicide safeners. This presentation summarizes several herbicide safener studies that were conducted by the Cutulle lab in sweetpotato (*Ipomea batatas*) and brassica leafy green crops. The majority of the natural products tested in these studies such as melatonin and brassinosteroids are believed to increase expression of cytochrome P450 enzymes, glutathione *S*-transferases or glycosyl transferases, which are involved in detoxification of xenobiotic materials. Though some of the compounds act as an anti-oxidant solely or in combination with induction of detoxification enzymes. Based on the screening of multiple herbicides in multiple bioassays ranging from tissues culture to field studies in sweetpotato and brassica leafy greens it has been determined that tank mixed melatonin with brassinosteroids appeared to be the most effective treatment for reducing injury from the herbicide bentazon on sweetpotato. The best safener herbicide treatment combination for brassica leafy greens was melatonin seed treatments on collard greens (*Brassica oleracea*)

95
Herbicide Carryover Trials: The Search For NS.

M. Vangessel[1], B. Scott[1], K. Vollmer[2], R. Batts[3]

[1]University of Delaware, Georgetown, United States, [2]University of Maryland, Queenstown, United States, [3]IR-4 Project, Raleigh, United States

Rotational restrictions from herbicide applications often limit vegetable producers' ability to diversify weed management strategies, increasing the risk of herbicide resistance. Lengthy rotation intervals may also force growers to plant crops with lower profit the following year instead of highly profitable crops. This study evaluated the feasibility of reducing labeled rotational intervals for three herbicides (pyroxasulfone, topramezone, and metribuzin) on five vegetable crops commonly grown in the mid-Atlantic region: processing green peas ('Jumpstart'), pickling cucumbers ('Expedition'), snap beans ('Caprice'), lima beans ('Cypress'), and seedless watermelon ('Fascination'). Field trials were conducted on loamy sand soils in 2023 and 2024 at the University of Delaware's Research and Education Center near Georgetown, DE. Herbicide rates (1x, 2x, and 4x the labeled rate) were applied the preceding summer, followed by vegetable planting the next growing season. Pyroxasulfone, topramezone, and metribuzin 1X rate was 128, 49, 262 g ai /ha^{-1}, respectively. Crops were planted "early" when initial growth rate was expected to be slower for the respective crop. Crop performance was assessed through stand counts, visual injury ratings, and yield measurements.
Green peas exhibited high tolerance to all herbicides, with no injury or yield effects. Watermelon showed minor early-season stunting (≤13%) with all herbicides, but yields were unaffected. Snap beans and lima beans experienced stunting in 2023 with metribuzin and topramezone, though no yield reductions were observed. In cucumbers, metribuzin and topramezone caused stunting of up to 23% at the 4x rate, but yields remained stable.
A companion study with University of Maryland (Wye Research Center, Queenstown, MD) evaluating metribuzin (1x and 2x rates) on additional crops (spinach, sweet potato, watermelon, snap beans, and lima beans) corroborated these findings, showing no significant injury across both locations.
Results suggest that rotational intervals for these herbicides can be reduced for specialty crops grown in the mid-Atlantic, offering farmers greater flexibility in crop rotations and weed management. Herbicide manufacturers are encouraged to consider the long-term benefits of shorter rotational intervals for vegetable growers and support label revisions to provide more options to growers, when supported by sound data.

96

Locked and Loaded: Abrasive Weed Management in Potato

A. McKenzie-Gopsill[1], M. J. Simard[2], R. Nurse[3], J. Bae[4]

[1]*Agriculture and Agri-Food Canada, Charlottetown, Canada*, [2]*Agriculture and Agri-Food Canada, Saint-Jean-sur-Richelieu, Canada*, [3]*Agriculture and Agri-Food Canada, Harrow, Canada*, [4]*Agriculture and Agri-Food Canada, Agassiz, Canada*

Abrasive weed control is a novel weed management tactic that uses air-propelled grit material directed at weeds with the intention of inflecting lethal damage. Over the past three years we have evaluated the use of various abrasive grit materials as a component of an integrated weed management program for use in the potato cropping systems of Atlantic Canada. Tested abrasive grits have included crushed walnut shells and corn cobs as well as sodium bicarbonate applied to emerged weeds once prior to potato flowering and again two weeks after. In the first two-years of the experiment the second abrasive grit was followed up with an application of rimsulfuron whereas the third year a combination of caprylic and capric acid was applied with a hooded sprayer to the furrows. In each year of the experiment, weeds of select species were tagged prior to treatment and monitored throughout the season. Neither the use of walnut or corn grit decreased biomass of weed species compared to the control yet while not controlled these species failed to produce seed. In contrast, sodium bicarbonate gave consistent and high levels of control of annual species including common lambsquarters (*Chenopodium album*), corn spurrey (*Spergula arvensis*) and common dandelion (*Taraxacum officinale*) across years. Potato marketable yield was not impacted by abrasive grit material but was higher following the use of either herbicide. Our results show that abrasive grits and in particular sodium bicarbonate can be incorporated into a traditional potato production to reduce seedbank inputs without impacting yield.

97

Tolerance of Rooted Sweetpotato (Ipomoea batatas) Transplants to Herbicides

S. Crawford[1], K. Jennings[1], D. Monks[1], D. Jordan[1], J. Schultheis[1], A. Gorny[1], L. Moore[1], K. Goble[1], H. Thome[1]

[1]*North Carolina State University, Raleigh, United States*

Sweetpotato (*Ipomoea batatas*) is commercially established with the use of nonrooted vegetative cuttings transplanted into the field from slip beds. However, limited research has explored the use of greenhouse grown transplants. This technique allows cuttings to develop roots and become more established prior to field transplantation. Using rooted transplants has the potential to increase herbicide tolerance. Field studies were conducted at the Horticultural Crops Research Station in Clinton NC in 2024 to evaluate the herbicide tolerance on greenhouse grown rooted transplants vs traditionally grown transplants (nonrooted cuttings) on sweetpotato. Treatments were arranged in a 2 X 6 factorial design with the first factor being holding time and the second factor herbicide. Greenhouse grown transplants were established in the greenhouse two wk prior to transplanting. Herbicide treatments included fluridone at 336 g ai ha^{-1}, flumioxazin at 107 g ha^{-1} plus sulfentrazone at 210 g ai ha^{-1}, flumioxazin PRE at 107 g ha^{-1} fb pyroxasulfone at 292 g ai ha^{-1}, flumioxazin PRE at 107 g ai ha^{-1} fb linuron at 420 g ai ha^{-1}, flumioxazin PRE at 107 g ai ha^{-1} fb metribuzin at 315 g ai $ha\text{-}^{-1}$, and a nontreated check for comparison. The study was maintained weed-free through cultivation and hand weeding as necessary. Flumioxazin fb metribuzin and sulfentrazone plus flumioxazin decreased total yield while fluridone, flumioxazin fb linuron, and flumioxazin fb pyroxasulfone did not have affect total yield. Increased culls were observed from the containerized sweetpotato transplant treatment. Inversely, no. 1 storage roots was greater for traditionally planted sweetpotato.

98
The Potato Vine Crusher: A New Tool for Harvest Weed Seed Control

A. McKenzie-Gopsill[1], A. N. MacDonald[1], L. Anderson[1], S. White[2], C. Noronha[1]

[1]*Agriculture and Agri-Food Canada, Charlottetown, Canada*, [2]*Dalhousie University, Truro, Canada*

Harvest weed seed control (HWSC), an evolving strategy in weed management, is highly effective for control of a variety of weed species in North American cropping systems. Previous devices for weed seed devitalization at harvest have been limited to tow-behind and integrated combine systems. The potato vine crusher (PVC) is a harvester-mounted set of rollers originally designed for crushing and control of *Ostrinia nubilalis* (Hübner) larvae during potato (*Solanum tuberosum* L.) harvest. To evaluate potential of the PVC for HWSC, we conducted stationary testing of spring tension and roller speed settings to maximize devitalization of lambsquarters (*Chenopodium album* L.), the most problematic weed species in Canadian potato production. In addition, we evaluated efficacy of the PVC for devitalization of several pernicious weed species under controlled conditions and during a simulated harvest. Increasing PVC spring tension reduced devitalization of lambsquarters seed, whereas roller speed had minimal effect. In contrast, maximized spring tension and minimized roller speed reduced lambsquarters emergence (53%) in soil. Hypocotyl and radicle elongation was observed from lambsquarters seed fragments under controlled conditions, potentially contributing to increased control in soil through fatal germination. High levels of seed devitalization (65 – 94%) was observed for all species under controlled conditions. During simulated harvest, control of large weed seeds (50 – 63%) was observed whereas smaller seeds were not impacted signifying the importance of seed size for PVC efficacy. These studies demonstrate the PVC as a promising new tool for HWSC in Canadian potato production systems.

99
Metamitron Efficacy and Safety in Table Beets

L. Sosnoskie[1], T. Besançon[2]

[1]Cornett AgriTech, Geneva, United States, [2]Rutgers University, New Brunswick, United States

Table beets are extremely sensitive to weed competition due to their slow and uneven germination and growth. Weed suppression is essential to reduce the intensity of interspecific interactions throughout the season, maximize yield quantity and quality, and improve harvest efficiency. The number of active ingredients registered for use is table beets is limited; additionally, most have narrow spectrums of weed control. Torero (Adama) is a pre-mix of metamitron (WSSA 5) and ethofumesate (WSSA 15), registered for PRE and POST use in beets, internationally. The inclusion of Torero in North American production systems adds an extra active ingredient (metamitron) to a very limited chemical toolbox.
In 2023 and 2024, research trials were conducted to evaluate the performance and safety of Torero in 'Red Cloud' beets at the Cornell AgriTech vegetable research farm in Geneva, NY (Honeoye loam), and at the Rutgers research station in Bridgeton, (NJ (Chillum silty loam). Treatments included: a nontreated check, S-metolachlor (Dual Magnum) at 0.72 kg ai ha-1 plus ethofumesate (Nortron) at 0.39 kg ai ha-1 PRE followed by (fb) a grower standard POST, metamitron (plus ethofumesate as Torero) at 1.64, 2.46 and 3,27 kg ai ha-1 PRE fb a grower standard POST, S-metolachlor (Dual Magnum) at 0.72 kg ai ha-1 plus metamitron (plus ethofumesate as Torero) at 1.64, 2.46 and 3,27 kg ai ha-1 PRE fb a grower standard POST, metamitron (plus ethofumesate as Torero) at 1.14 kg ai ha-1 fb metamitron (plus ethofumesate as Torero) at 1.14 kg ai ha-1, and S-metolachlor (Dual Magnum) at 0.72 kg ai ha-1 plus metamitron (plus ethofumesate as Torero) at 1.14 kg ai ha-1 fb metamitron (plus ethofumesate as Torero) at 1.14 kg ai ha-1. Ethofumesate rates ranged from 0.49 kg ai ha-1 to 1.44 kg ai ha-1 for the Torero applications. Weed cover, weed control, and crop injury data were collected throughout the season.
At both sites, Torero programs reduced weed cover/improved weed control, relative to the nontreated check) as effectively as the grower standard (Dual Magnum plus Nortron fb Upbeet, Stinger, and/or Spin-Aid). Control of common lambsquarters (Chenopodium album), redroot pigweed (Amaranthus retroflexus), purslane (Portulaca oleracea), and Polygonum spp. ranged from 90-100%. Common ragweed (Ambrosia artemisiifolia) densities were not sufficient to run statistical analyses, however, PRE applications of Torero at the higher rates were more effective than Dual Magnum plus Nortron at controlling the species. Beet injury in response to Torero was greater in NJ compared to NY. On the coarser soils (NJ), stunting ranged from 5-45% 14-28 days after PRE applications; In NY, stunting did not exceed 16%. PRE injury tended to be greatest at the highest rates and where Dual Magnum was included as a tank mix partner. POST injury was not observed at either site. Despite the observed early season injury, crop yields were not impacted. All Torero applications significantly improved marketable beet yields over the nontreated check. Results suggest that Torero is effective for weed control in Eastern US beets. Soil conditions will enhance injury; use on sandy soils will not be advised.

Gap in pagination due to unavailable paper.

Page 100

101

Detection and Mapping of Carolina Redroot in Cranberry Bogs using Deep Learning and Drones.

T. Nguyen[1], H. Nguyen[1], D. Ham[1]

[1]*Rowan University, Glassboro, United States*

In this talk, we present a framework for the detection and mapping of Carolina redroot (*Lachnanthes caroliana*) in cranberry bogs. Our goal is to build a smart drone that can fly autonomously above cranberry bogs, detect Carolina redroot, and map out the locations of the weed or perform spot spraying. The proposed system employs a deep learning-based Carolina redroot detection model which is trained on a manually annotated dataset of high-resolution visual images of cranberry bogs taken by drones. An autonomous drone is programmed to fly over a cranberry bog to collect images which are then sent through the trained deep learning model which detects the locations of the weed. The ground locations of the detected Carolina redroot are computed using a linear image-to-ground transformation. The spatial mapping capabilities of the system help farmers to reduce herbicide usage and minimize environmental impact.

102
Evaluating Unmanned Aerial Systems for spraying herbicides

V. Singh[1], F. Esmaeilbeiki[1], A. S. Brar, R. Wamanse[1], M. Viric[1], D. Martin[2]

[1]*Virginia Tech, Painter, United States*, [2]*United State Department of Agriculture, College Station, United States*

The integration of Unmanned Aerial Systems (UAS) into agricultural practices has opened new possibilities for precision crop management, particularly herbicide application for row-crops, and seeding of cover crops. Several studies were conducted to evaluate efficacy, and droplet spectra for preemergence (PRE) and postemergence (POST) herbicide applications. PRE and early POST applications with UAS provided weed control similar to backpack and tractor sprayer. However, UAS provided 20-25% greater weed control in case of late-POST applications. Greater efficacy with UAS at late-POST was attributed due to air circulation and higher concentration of herbicides applied. UAS generated air circulation provided greater distribution and coverage compared to tractor sprayers at late-POST applications.

103
Investigating the Naio OZ Autonomous Robotic Platform for Weed Control in Sweet Corn

T. Besançon, C. Plassmeyer[1], W. Bouchelle[2], L. Sosnoskie[3], M. Bertucci[1]

[1]*University of Arkansas, Fayetteville, United States*, [2]*Rutgers University, New Brunswick, United States*, [3]*Cornell University, Geneva, United States*

Weed management in vegetable crops is challenging due to limited herbicide options, differential cultivar sensitivity to herbicide phytotoxicity, and low commercial tolerance for crop injury. Labor shortages, increased wages, and herbicide-resistant weeds further complicate management strategies. Robotic weed control technologies may address these challenges by reducing herbicide dependence, increasing weeding precision, and enabling earlier weed detection and removal. This study evaluated the performance of a row-following autonomous Naio OZ robot in sweet corn compared with standard preemergence (PRE) and postemergence (POST) herbicide applications. The Naio OZ is a compact (150 kg), fully electric autonomous seeder and mechanical weeder that uses real-time kinematic (RTK) GPS correction for precise operations. Field trials were conducted in 2024 in New Jersey using a randomized complete block design with four replicates on single-row sweet corn planted 81 cm apart. Treatments included: weekly or biweekly mechanical weeding with the OZ robot starting 7 days after planting (DAP); *S*-metolachlor at planting (1,815 g ai ha^{-1}) alone or followed by tembotrione plus atrazine 30 DAP (92 and 1,680 g ai ha^{-1}, respectively) or biweekly mechanical control starting 21 DAP; and untreated and weed-free controls. The Naio OZ was fitted with three 20-cm A-blades and two 23-cm finger weeders on a trailing toolbar. Weed density between crop rows 8 weeks after planting was significantly lower following PRE, PRE followed by POST, and weekly OZ cultivation compared to untreated controls. However, OZ cultivation (weekly or biweekly) did not reduce within-row weed density compared to untreated controls. Weekly OZ cultivation provided similar weed biomass reduction (73%) as PRE followed by POST (61%) compared to untreated controls. PRE alone, biweekly OZ cultivation, or mixed strategies did not significantly reduce weed biomass (23% reduction) compared to untreated controls. Yield was not significantly reduced with PRE followed by POST, whereas other treatments reduced sweet corn yield 40% (weekly OZ) to 58% (biweekly OZ) compared to weed-free controls. Without weed control, yield decreased 65%. The finger weeders were ineffective for in-row cultivation, causing significant yield reduction despite acceptable between-row weed control. These results demonstrate that while the Naio OZ robot effectively controlled weeds between rows, finger weeder attachments were unsuitable for in-row cultivation in sweet corn. Future research should evaluate alternative toolbar attachments that improve in-row cultivation without causing crop lodging.

104

Innovative Approaches to Weed Control: Leveraging Laser Weeding Technology in New Jersey Spinach and Pea Crops

T. Besançon, W. Bouchelle[1], L. Sosnoskie[2]

[1]Rutgers University, New Brunswick, United States, [2]Cornell University, Geneva, United States

Limited herbicide options, resistance development, crop injury risks, and regulatory changes are driving interest in non-chemical weed management alternatives for processing vegetables. In 2024, field trials in New Jersey evaluated laser weeding efficacy using a Carbon Robotics LaserWeeder™ in spinach (*Spinacia oleracea* L.) and pea (*Pisum sativum* L.). **Spinach treatments** included: preemergence (PRE) *S*-metolachlor (534 g ai ha^{-1}) alone or followed by laser weeding (3 passes) or postemergence (POST) herbicides (phenmediphan 547 g ai ha^{-1} plus clethodim 118 g ai ha^{-1}); POST herbicides alone; and weedy and weed-free controls. **Pea treatments** included: laser weeding alone (2 passes), POST herbicides (pyridate 790 g ai ha^{-1} or bentazon plus imazethapyr 1,150 + 50 g ai ha^{-1}), and weedy and weed-free controls. Laser weeding equaled or exceeded PRE and POST herbicides in controlling common lambsquarters (*Chenopodium album* L.) and common ragweed (*Ambrosia artemisiifolia* L.). Compared to untreated controls, laser weeding reduced total weed cover and density, resulting in ≥97% reduction in end-of-season weed biomass. In pea, POST herbicides caused greater crop stunting at harvest (>10%) than laser weeding (0%). Pea dry biomass was reduced 23–28% following POST herbicide applications and 38% in weedy controls compared to weed-free controls, with no significant reduction following laser weeding. Shelled pea yield declined 9–25% in POST herbicide-treated plots but showed no reduction following laser weeding. In spinach, laser weeding (3 passes) reduced weed dry biomass 99.9% compared to 70% reduction following herbicide applications. Greater spinach stunting (≥8%) occurred in plots receiving *S*-metolachlor PRE regardless of POST treatment compared to laser-weeded plots. Spinach head weight was 39% higher in laser-weeded plots than in plots receiving *S*-metolachlor PRE followed by laser weeding POST, despite similar weed control efficacy. These results demonstrate that multiple laser passes effectively control weeds without phytotoxicity, yielding superior crop performance compared to conventional herbicides. Further research should optimize laser weeding parameters across diverse environments and weed species and evaluate commercial units with enhanced laser capacity and processing speed.

105
Weed and Soil Biological Responses to Zasso Electrical Weeding.

A. Butler-Jones, E. Maloney[1], J. Owens[2], G. Peck[2], K. Wickings[1], B. Hanson[3], M. Moretti[4], L. Sosnoskie[1]

[1]Cornell University, Geneva, United States, [2]Cornell University, Ithaca, United States, [3]University of California, Davis, Davis, United States, [4]Oregon State University, Corvallis, United States

The continued evolution of herbicide resistant weeds and growing consumer demand for non-chemical-based weed management have sparked grower interest in novel technologies including electrical weed control (EWC). Despite this growing interest in EWC, there is limited information available regarding its impact on weed and soil biological communities. The objective of this research is to evaluate the impact of electrical weeding application parameters on weed cover, weed biomass, soil microarthropod abundance, and soil microbial respiration. To accomplish these objectives, research trials were conducted in weedy, unplanted agricultural fields in 2023 at Cornell AgriTech in Geneva, NY. A tractor-mounted and PTO-driven Zasso Electroherb unit was operated in the fields at two travel speeds (1.9 km/h, 4.8 km/h) and three mean amperage settings (9.1 Amps, 19.5 Amps, 35.5 Amps). The experiment was organized as a randomized complete block design (RCBD) with four replicates and two experimental runs. Dominant weed species across sites included common lambsquarters (*Chenopodium album* L.), Powell amaranth (*Amaranthus powellii* S. Wats.), common ragweed (*Ambrosia artemisiifolia* L.), large crabgrass [*Digitaria sanguinalis* (L.) Scop.], yellow foxtail [*Setaria pumila* (Poir.) Roem. & Schult.], and yellow nutsedge (*Cyperus esculentus* L.). Percent weed cover was estimated using Canopeo, a smartphone application that measures percent green area, within 24 hours of treatment and up to 19 days after treatment (DAT). Additionally, weed biomass was collected 28 DAT. Soil samples were taken 1 DAT and used to assess the impact of EWC on microarthropod abundance and microbial respiration. When averaged across the season, we observed significant differences in percent weed cover based on tractor speed and unit amperage setting ($P < 0.05$). In addition, EWC reduced total weed biomass 28 DAT by 84% to 88% compared to the non-treated check ($P < 0.05$); however, broadleaf weed biomass did not significantly differ between the EWC plots. No differences in total microarthropod abundance, collembola abundance, or the ratio of mites:collembola were observed between non-treated and EWC plots; however, electricity applied at 1.9 km/h with a mean amperage of 35.5 Amps significantly reduced total mite and oribatid mite abundance compared to the non-treated check. Further, no difference in carbon dioxide (CO2) flush was observed between treatments; however, the impact of EWC on microbial respiration over time is currently being assessed. Trials will be repeated in 2025.

106
Grape sucker control with optically-guided sprayer

T. Besançon[1], L. Sosnoskie[2]

[1]*Rutgers University, New Brunswick, United States*, [2]*Cornell University, Geneva, United States*

Optically-guided, targeted spray systems show promise for reducing herbicide use, minimizing crop damage, and expanding weed control options in perennial crops. In grapevines, suckers (basal non-productive shoots) divert nutrients from desirable tissues, altering fruit production and development. Unmanaged suckers also increase tissue available for pest and pathogen colonization and impede production practices and harvest. Traditional management involves manual removal or targeted spraying, which presents labor challenges, costs, and risks of off-target herbicide movement.
Field trials conducted in New York and New Jersey (2022, 2024) evaluated the efficacy of the WEED-IT™ precision spraying system for controlling grape suckers. The WEED-IT™ emits blue light (450-500 nm) that induces chlorophyll fluorescence, producing near-infrared signals that sensors detect to differentiate living plants from surroundings and activate targeted herbicide application.
In 2022, a New York trial compared continuous directed application using a backpack sprayer versus targeted application using the WEED-IT™ system. Herbicides evaluated were carfentrazone (35 g ai ha^{-1}), tiafenacil (105 g ai ha^{-1}), and glufosinate (1,190 g ai ha^{-1}). No differences in sucker control efficacy were observed between systems, while the WEED-IT™ system reduced herbicide use by 40%.
In 2024, trials in New York (Concord) and New Jersey (Cabernet Franc) compared the WEED-IT™ system to spray wand applications and manual removal. Carfentrazone (35 g ai ha^{-1}) was applied in both locations, while tiafenacil (105 g ai ha^{-1}) was used only in New Jersey. At 21 days after treatment (DAT), sucker control in New York was 98% with the WEED-IT™ system, 94% with the spray wand, and 90% with manual removal. In New Jersey, the WEED-IT™ system achieved 84% control compared to 67% for manual removal. Herbicide applications (spray wand or WEED-IT™) required approximately 40 seconds per 100 ft of row, compared to 120 seconds for manual removal. Averaged across herbicides, sucker fresh biomass suppression with the WEED-IT™ system (-92%) was comparable to wand application (-94%) and more effective than hand removal.

107
Performance and Safety of Electrical Weeding in Annual Crops

L. Sosnoskie[1]

[1]*Cornell AgriTech, Geneva, United States*

Many growers are interested in identifying novel weed control practices that are effective against herbicide resistant species, limit soil disturbance and reduce or eliminate the need for synthetic herbicides. One such technology is electrical weeding. Electrical weeding is an old technology (first patents issued in the 1800's) that has seen renewed interest in US agriculture. Electrical weeding controls unwanted plants by applying an electric current directly to leaves and stems. The flow of electricity through the plant generates heat, which disrupts cells and damages tissues. In New York, there are opportunities to engage in electrical weeding. These include:

a) herbicide resistance management in horseweed (*Erigeron canadensis*), Palmer amaranth (*Amaranthus palmeri*) and waterhemp (*A. tuberculatus*) in soybeans, specifically to reduce competing vegetation at harvest and prevent seed set;
b) Applications in snap beans, beets, cabbage because of limited herbicides, potential for PRE herbicide failure, and because these are shorter stature crops that allow for height differentials between crops and many troublesome weeds such as lambsquarters (*Chenopodium album*), pigweeds (*Amaranthus* spp., and common ragweed (*Ambrosia artemisiifoli*a);
c) Use in a growing organic industry with relatively limited weed control options, particularly late season.

Observational studies and repeated research trials were conducted in the fields of grower-collaborators (soybean and beets) in 2020 and 2021. All applications were made using Weed Zapper electric weeders made by Old School Manufacturing. Results found that lambsquarters, pigweed species, and common ragweed could be controlled >60% by electrical weeding if applied when plants were succulent and actively growing. Grass weeds and velvetleaf (*Abutilon theophrasti*) were more difficult to manage. Older, "tougher" and perennial weeds were also less well controlled by electrical weeding. Species whose heights did not exceed that of the canopy, like Galinsoga spp., were also not controlled.

Seed production by lambsquarters was significantly reduced when timely electrical weeding occurred (and plants were severely damaged). In some instances, only the tops of velvetleaf plants (tissue above the contact point with the bar) were injured by electrical weeding. Weed seed that formed in capsules produced on damaged/wilted stems and branches were smaller in size (17.8 g seed per 100 capsules) and did not germinate as well (5% germination) as seed produced in capsules formed on healthy tissue located below the contact point (42.2 g per 100 capsules and 95% germination). Soybean injury was minimal across all studies conducted in that crop. With respect to beets, damage (necrotic tap roots, production of compensatory feeder root, leaf wilting and death) was occasionally observed in plants directly adjacent to "zapped" weeds. Occurrences were often related to operations conducted on the highest unit setting or under dry soil conditions. In the latter scenario, it was speculated that the current took the path of least resistance through the (relatively wet) beet roots. Electrical weeding can be effective at controlling some weed species, such as erect, annual broadleaves. More work is needed to better understand the conditions resulting in beet injury.

108

Tillage intensity effects on weed recruitment dynamics in organic soybean.

A. Hamilton[1], J. Wallace[1], M. Barbercheck[1], M. Bilenky[2]

[1]Penn State, University Park, United States, [2]Purdue University, West Lafayette, United States

Opportunities for intensifying grain production on organic farms include (1) reducing yield-limiting factors (i.e., increase yield/ac) through improved weed, pest, and nutrient management practices, 2) increasing operational efficiencies to permit expansion of land in production (i.e., increase acres/farm), and 3) simplifying rotations by reducing use of perennial forages (i.e., increase grain acres/year). Each of these intensification strategies must be coupled with conservation practices that maintain or improve environmental quality for sustainable intensification goals to be realized. We conducted a three-year, cropping system experiment that quantified tradeoffs among provisioning (crop yield), regulating (pest control, nutrient balance), and supporting (soil health, soil erosion) services that contribute to sustainable intensification. We tested three experimental systems that used alternative cover cropping and reduced-tillage practices to produce a tillage intensity, tillage frequency, and perenniality (i.e., length of time without tillage) gradient within a 3-yr corn - soybean -wheat rotation using a full-entry experimental design. The 'NT cover crop' experimental system was designed to represent tillage practices comparable to commercial feed grain rotations for grain crops in the Northeast while implementing reduced-tillage practices for establishing cover crops. Inversion (moldboard plow). was used preceding corn and soybean. Relay cover cropping practices were employed to reduce tillage intensity and frequency within the rotation. The 'shallow-tillage' experimental system was designed to quantify the effects of emerging shallow, non-inversion tillage practices (high-speed disk) as a primary tillage tool in a corn – soybean – wheat rotation. The 'rotational no-till soybean' experimental system was designed to integrate no-till soybean planting and relay cover cropping practices at appropriate phases in the rotation. Soybean is no-till planted into roll-crimped cereal rye. Multiple indicators of crop performance, pest management performance, soil health, and other ecosystem services were measured throughout the cropping system experiment. Soybean sequence results are shared in this presentation. Our results showed that use of shallow non-inversion tillage resulted in significantly greater tillage intensity, measured using the STIR index, in comparison to experimental systems that employed either no-till cover crop or cash crop sequences. Soybean yields did not differ among experimental systems, but net income from the cover crop to soybean sequence was significantly greater in the rotational no-till soybean sequence due to lower input costs. Sequences using rotational no-till soybean or shallow non-inversion tillage resulted in greater late season weed biomass compared to the soybean sequence that employed inversion tillage. However, a trend towards lower active carbon in the inversion tillage sequence compared to other sequences was observed, suggesting that primary tillage and cultivation passes came at soil carbon cost. These results demonstrate that organic no-till soybean is viable from a production and economic perspective but creates additional weed management challenges within an organic rotation.

109

Erigeron canadensis tolerance to herbicides and climate change: a global assessment

I. Moisinho[1], A. Witter[2], D. Nagy[3], C. Rosche[4], C. Brunharo[1]

[1]The Pennsylvania State University, University Park, United States, [2]Maringa State Universiy, Maringa, Brazil, [3]Martin-Luther-University Halle, Halle, Germany, [4]Martin-Luther-University Halle, Haale, Germany

Understanding the physiological responses of weeds to environmental changes and herbicides is important for planning weed management strategies and anticipating weed behavior under future climates. We aimed to evaluate the tolerance of *Erigeron canadensis* populations under conditions of increased temperature and sub-lethal herbicide exposure. We utilized approximately 500 populations from 34 countries around the world obtained from the iConnect Network. Populations were pre-acclimated to 25°C (12h daylength) for two days prior to imposing heat stress for seven days at 36°C. Chlorophyll *a* fluorescence measurement (F_v/F_m) were recorded after acclimation and seven days after heat stress initiated. Metribuzin stress was induced by using a sub-lethal rate (35 g ai ha^{-1}) and F_v/F_m were taken zero and two days after treatment. Populations with greater heat stress tolerance originated from regions with high temperatures, low precipitation and high radiation. A latitudinal gradient in the response to oxidative stress caused by metribuzin was observed. In addition, native populations from higher latitudes are less herbicide stressed and non-native populations were overall less herbicide stressed. Plants experiencing higher stress levels were from locations with elevated annual mean temperatures and radiation. Physiological mechanisms to tolerate heat and metribuzin stress seem to be different. Understanding the mechanism involved in *E. canadensis* adaptation could improve management, delay herbicide-resistance evolution and develop strategies to minimize the consequences of climate change. Future work will focus on genome-wide association studies with to identify the genetic variants associated with oxidative stress tolerance and glyphosate resistance. We are currently phenotyping populations to elucidate the genetic controls of germination dynamics and vernalization requirement.

Gap in pagination due to unavailable paper.

Page 110

111

Evaluation of a Machine-Vision Sprayer for Weed Control in Managed Turfgrass

S. Askew[1], J. McCurdy[2], A. Wilber[2], J. Romero[1], N. Godara[1], P. Le Naour Vernet[3]

[1]Virginia Tech, Blacksburg, United States, [2]Mississippi State University, Starkville, United States, [3]Ecorobotix, Yverdon-les-Bains, Switzerland

Advancements in machine learning have facilitated on-demand weed identification and targeted treatment, leading to the commercialization of "see and spray" agricultural sprayers globally. Although machine-vision technology has been extensively explored in production agriculture, its application in turfgrass management is underexplored. The ARA, a machine-vision sprayer from Ecorobotix, initially developed primarily for vegetable crops, is now being adapted for use in ornamental turf. This study investigates the performance of the vegetable-crops version of the ARA, which has limited training on athletic-height Kentucky bluegrass turf, in controlling thin paspalum (*Paspalum setaceum*) on both creeping bentgrass golf fairways and adjacent tall fescue and Kentucky bluegrass golf roughs. Two randomized complete block experiments were conducted to evaluate the ARA's accuracy and its impact on pesticide use reduction. The results demonstrated significant reductions in herbicide application, with over 90% less chemical used on creeping bentgrass fairways and more than 70% reduction on tall fescue roughs, translating to potential savings of up to $500 per hectare. Digital analysis via drone imagery revealed that the machine vision sprayer treated 4.5% of the total area on creeping bentgrass and 25.7% on tall fescue. The machine-vision sprayer accurately treated 88% of the 238 thin paspalum plants encountered on the golf rough and 100% of broadleaf weeds. Of over 1000 goosegrass plants assessed in the rough, only 20% were treated, possibly due to the sprayer's limited turf-specific training and the similarity between goosegrass leaf blades and the surrounding turf. The weed-targeting accuracy was notably lower on the creeping bentgrass fairway than on the rough, underscoring the influence of training data on performance. These findings align with previous studies showing that while machine vision sprayers can offer commercially viable accuracy, their effectiveness largely depends on the quality and specificity of the training data. This suggests there is considerable potential for enhancing weed management practices in turfgrass settings through further development and training of such technologies.

112

Are Directed Applications of POST and PRE Herbicides Feasible in Container Grown Nursery Crops?

J. Neal[1], C. Harlow[1]

[1]*North Carolina State University, Raleigh, United States*

The lack of selective postemergence herbicide options for container-grown nursery crops results in a reliance on hand-removal of weeds. Yet, increasing costs and limited availability of labor is a significant concern for growers. Directed applications of PRE and POST herbicides, avoiding tender foliage, have been a common practice in field-grown horticultural and agronomic crops. However, this application technique is impractical in container-grown nursery crops due to the low canopy height and close plant spacing. In this study, a sprayer capable of directed sprays in containers was constructed and tested. A commercial potter's wheel was retrofitted with sensors and an Arduino processing board and software to control rotational speed and on/off spray cycles synchronized with the rotation. Standard 4-L plastic pots were turned at a constant speed, and a UB8501 off-center spray nozzle was positioned 5 cm above the substrate surface. The prototype sprayer produced a uniform spray application to the substrate surface and beneath the canopy of commonly grown nursery crops. Operating parameters for these tests included one UB8501 nozzle, 14.5 psi, 60 rpm rotation, and spray duration of 1 or 2 rotations (=1 or 2 seconds), providing acre-equivalent spray volumes of 174 or 350 GPA. *Hydrangea macrophylla*, *Thuja* 'Green Giant' and *Rosa* x Radtko 'Knockout Double Red' liners were potted August 12, 2021. Additional pots were filled with potting substrate and seeded to *Cardamine flexuosa* or *Euphorbia maculata*. On September 3rd, preemergence herbicides (flumioxazin, indaziflam and isoxaben + prodiamine) were applied at labeled doses over the top of plants with a backpack sprayer or as directed sprays with the prototype sprayer. In a separate experiment on September 17th (~5 weeks after potting), diquat, glufosinate and flumioxazin were applied as directed sprays with the prototype sprayer to the same crop and weed species. Crop injury and weed control were visually assessed and above-ground fresh weights measured. Over-the-top applications of preemergence herbicides injured *Rosa* and *Hydrangea*, but directed applications did not. Similarly, directed applications of diquat, glufosinate or flumioxazin did not injure *Rosa* or *Thuja*. Directed applications of postemergence herbicides caused little reduction in visual quality of *Hydrange*a but reduced fresh weights by 27 to 50% compared to the nontreated. Weeds were controlled by directed applications of preemergence or postemergence herbicides. This research provides proof of concept, that directed applications of broad-spectrum preemergence and postemergence herbicides in container nurseries are possible. Directed applications could improve preemergence weed control by allowing for the use of broader spectrum herbicides and also reduce hand weeding labor by allowing the use of postemergence herbicides. Design challenges remain to improve the efficiency of system operation, particularly increasing the speed of throughput.

113
Annual bluegrass encroachment into tall fescue subjected to traffic

B. Park[1], J. Murphy[1]

[1]*Rutgers, The State University of New Jersey, New Brunswick, United States*

High traffic sports fields are prone to annual bluegrass (ABG; *Poa annua* L.) invasion. Annual bluegrass encroachment can be mitigated, in-part, by establishing turfgrass cultivars that have improved traffic tolerance. Among the objectives of our field trial was to evaluate ABG encroachment into tall fescue (*Lolium arundinaceum* [Schreb.] Darbysh.) subjected to traffic stress. Entries comprising the 2018 National Turfgrass Evaluation Program (NTEP) Tall Fescue Test were seeded on a loam soil in North Brunswick, NJ in September 2018. Traffic was applied during 2019 to 2023 as a strip across half of each plot using a combination of the Rutgers Wear Simulator, Cady Traffic Simulator, and a vibratory roller; the other half of each plot did not receive traffic at any point during the trial period. Visual evaluations of fullness of turfgrass canopy (FTC; 0 to 100% scale; 100% = complete turfgrass canopy) and ABG cover (0 to 100% scale; 100% = complete ABG cover) were taken in traffic and no traffic lots prior to traffic application (3 and 16 May, respectively) and after 84 machine passes (2 October and 9 November, respectively) during 2023. Data were analyzed as a 2 (no traffic and traffic) x 132 (entries) factorial strip-plot design. The cumulative effects of traffic applied during 2019 to 2022 resulted in greater ABG cover in trafficked plots (34%) compared to no traffic plots (2%) on 16 May 2023. Forty-eight entries had the least ABG cover on this rating date (under traffic stress) and Kentucky-31 had the greatest ABG cover in the absence of traffic (32%) and under traffic stress (88%). Traffic applied prior to 2023 did not affect ABG encroachment into GLX ACED, Finelawn Supreme, JT 233, Bullseye LTZ, Southside, Essential 2, and Topshelf; each of these entries were among those with the least ABG cover under traffic stress on 16 May 2023. Traffic applied during 2023 resulted in tall fescue plots having greater ABG cover (48%) compared to non-trafficked plots (20%) on 9 November 2023. ANOVA detected a significant traffic × entry interaction for FTC and ABG cover after traffic was completed in 2023. Ninety-three (93) tall fescue entries had the least ABG cover (30 to 50%) under the traffic condition on 9 November 2023; entries with the greatest FTC (25 to 45%) under traffic and least ABG cover under traffic were Monument, Topshelf, ATF 1768, GLX ACED, JT 233, PST-5TRN, Southside, Padre 2, Spyder 2LS, AST8118LM, PST-5THM, Bullseye, Lifeguard, OG-WALK, Birmingham, Naturally Green, ProGold, Bullseye LTZ, PPG-TF-249, Extravaganza, Moondance GLX, AST8218LM, Fairfield, TD2, SE5STAR, Providence, p Hemi, DLFPS-321/3703, BAR-FA8230, JT-517, Raptor III, K18-WB1, Hellcat GLR, Oriole, Capitan, Palomar, Gro-Pro, Grand Prix, Kizzle, Copious TF, DLFPS-321/3708, Stealth, and SETFM3. Entries with the least FTC and greatest ABG cover under traffic stress on 2 October and 9 November 2023, respectively, were Zion, Talladega II, RHF. PPG-TF-318, Bentley, Galactic, Tango, ATF2116, Bonfire, Dragster, DLFPS-321/3706, PST-5DZM, DLFPS-321/3694, Firehawk SLT, SR 8700, Barnoble, Kentucky-31, PST-5MINK, and Degas.

AUTHOR INDEX